Introduction to Sensors for Electrical and Mechanical Engineers

Introduction to Sensors for Electrical and Mechanical Engineers

Martin Novák

CRC Press
Taylor & Francis Group
Boca Raton London New York

CRC Press is an imprint of the
Taylor & Francis Group, an **informa** business

First edition published 2020
by CRC Press
6000 Broken Sound Parkway NW, Suite 300, Boca Raton, FL 33487-2742

and by CRC Press
2 Park Square, Milton Park, Abingdon, Oxon, OX14 4RN

ISBN: 978-0-367-51821-9 (hbk)
ISBN: 978-0-367-53401-1 (pbk)
ISBN: 978-1-003-08169-2 (ebk)

Typeset in LMRoman
by Nova Techset Private Limited, Bengaluru & Chennai, India

Visit the Taylor & Francis Web site at
http://www.taylorandfrancis.com

and the CRC Press Web site at
http://www.crcpress.com

To my family.

Contents

Preface

Why does an engineer need sensors?

Sensors are all around us. They are in phones, cars, planes, trains, robots, mills, lathes, packaging lines, chemical plants, power plants, etc. Modern technology could not exist without sensors. The sensors measure what we need to know, and the control system then performs the desired actions. When an engineer builds any machine he or she needs to have basic understanding about sensors. Correct sensors need to be selected for the design right from the start. The designer needs to think about the ranges, required accuracy, sensor cost, wiring, correct installation and placement, etc. Without the basic knowledge of sensors' fundamentals no machine can be built successfully today. On the other hand, the sensors are only one out of many steps of a successful design. No sensor or control system will help you if your machine is not designed properly from a mechanical point of view, if correct materials are not selected, or when there are any other flaws.

The purpose of this book is to provide basic knowledge about selected sensor technologies to engineers. The sensor principles will be discussed briefly, then their properties such as accuracy, range, linearity, etc., will be shown. Finally sensor and applications examples will be presented.

1

Measurement basics

1.1 General properties

Measurement is a very important topic in every technical system. If we cannot measure some physical property in our system (temperature, pressure, humidity, force, torque, liquid level, etc.) with a desired precision and other characteristics, we cannot expect to have good control over the system. It is therefore very important to know how we can measure the desired property, what sensor principles we can use, what accuracy can we expect, where to place sensors to obtain correct readings, etc. The proper application of any process-monitoring device requires a good understanding of how sensors work and their advantages as well as their limitations.

1.1.1 What is a sensor?

A sensor is a device that transfers the measured physical property into some output signal that can be processed or displayed. Today's sensors use mostly electrical output signals; however, some other possibilities exist like mechanical, pneumatic or hydraulic signals. The sensor is then usually placed in a measuring chain where we can find the sensor, signal processing (like filters, amplifiers, etc.) and gauges (voltmeter, ampere meter, etc.). This situation is shown in figure 1.1.

This arrangement is a classical sensor. We can find also so-called smart sensors, shown in figure 1.2, where all or some parts of the measuring chain are integrated in an independent block (can be integrated circuit, circuit board).

Smart sensors are most commonly composed of the main sensor (or sensors), amplifiers, analog/digital converters, microprocessor and some commu-

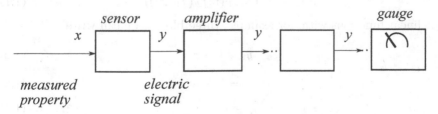

FIGURE 1.1: Block arrangement in a classical sensor

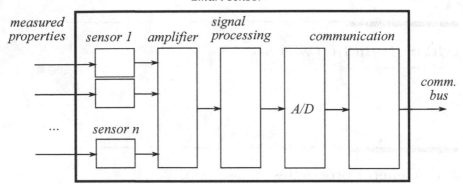

FIGURE 1.2: Block arrangement in a smart sensor

nication interface. In all those systems, the final measuring system behavior is given by properties of the individually connected blocks in the chain. By modifying individual block properties we can therefore change completely the system's behavior; e.g., we can make the sensor's non-linear characteristic linear.

1.1.2 Block algebra

Regardless of the type of arrangement, the whole-device properties are always given by the properties of the used blocks. Block algebra can be used to derive the properties of the whole device. Three possibilities of connecting the blocks will be discussed—serial, parallel and feedback connection. More complicated connections can be made by combining those simple approaches.

1.1.2.1 Serial connections

In a serial connection of blocks, the individual parts are connected so that the input of the next block is connected to the input of the preceding block. An example is shown in figure 1.3.

If the functions f_1, f_2,... ,f_n are known, the output signal y_n can be calculated as

$$y_n = f_n \left(f_{n-1} \left(...f_2 \left(f_1 \left(x \right) \right) \right) \right) \tag{1.1}$$

For linear functions with, for example, three blocks, we can write

$$y_1 = k_1 \cdot x \tag{1.2}$$

$$y_2 = k_2 \cdot y_1 \tag{1.3}$$

$$y = k_3 \cdot y_2 \tag{1.4}$$

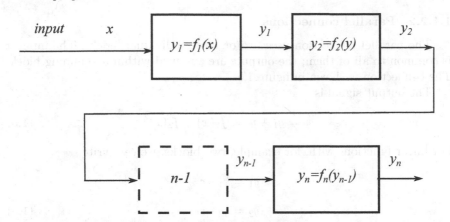

FIGURE 1.3: Block diagram for serial connection

The connection output will therefore be

$$y = k_1 \cdot k_2 \cdot k_3 \cdot x = K \cdot x \qquad (1.5)$$

where $K = k_1 \cdot k_2 \cdot k_3$.

The graphic solution can be found by simply plotting several charts in the same graph with the corresponding axes. An example is shown in figure 1.4.

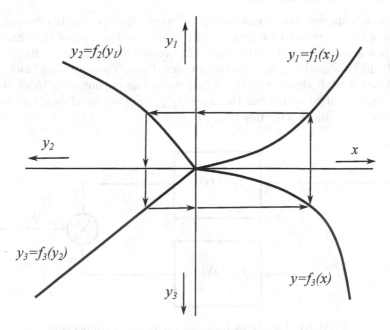

FIGURE 1.4: Graphic solution for serial connection

1.1.2.2 Parallel connections

The parallel connection uses two or more individual blocks. The input x is common to all of them; the outputs are summed within a summing block. The connection is shown in figure 1.5.

The output signal is

$$y = y_1 + y_2 = f_1(x) + f_2(x) \tag{1.6}$$

For linear functions with, for example, two blocks, we can write

$$y_1 = k_1 \cdot x \tag{1.7}$$

$$y_2 = k_2 \cdot x \tag{1.8}$$

The connection output will therefore be

$$y = k_1 \cdot x + k_2 \cdot x = (k_1 + k_2) \cdot x = K \cdot x \tag{1.9}$$

where $K = k_1 + k_2$.

The graphical solution (for any function, linear or non-linear) can be obtained in a single graph, with a common x-axis for all plotted curves. An example is shown in figure 1.6. The curves have a common x-axis. The output characteristic is obtained as a sum of individual characteristics.

1.1.2.3 Feedback connections

This is a special, very common type of connection in control systems. The feedback can be positive or negative depending on the sign of the signal fed back in one node from the other one. The connection is shown in figure 1.7.

The input signal is x, and the output signal is y. The summing block sums the signal x with signal x_1. The signal x_1 is the output of a block in the feedback, described with a function $x_1 = f_2(y)$. The summed signal x_2 is then fed back to the block with function $y = f_1(x_2)$.

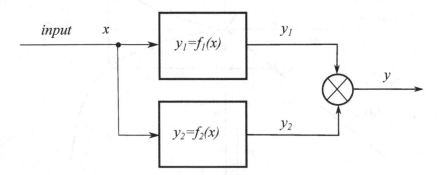

FIGURE 1.5: Block diagram for parallel connection

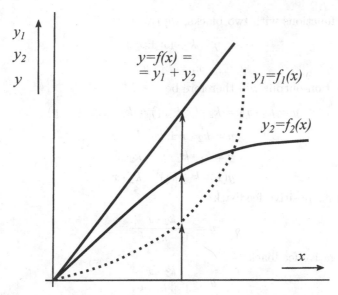

FIGURE 1.6: Graphic solution for parallel connection

The sign in the feedback sets the feedback type, either positive (sign $+$) or negative (sign $-$) feedback.

The connection output is

$$y = f(x \pm x_1) \tag{1.10}$$

For positive feedback

$$y = f(x + x_1) \tag{1.11}$$

and for negative feedback

$$y = f(x - x_1) \tag{1.12}$$

The negative feedback is more common in control systems as it improves system stability; positive feedback is in general not desired as it makes the system unstable.

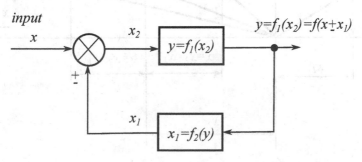

FIGURE 1.7: Block diagram for feedback connection

For linear functions with two blocks, we can write

$$y = k_2 \cdot (x \pm x_1) \tag{1.13}$$

$$x_1 = k_1 \cdot y \tag{1.14}$$

The connection output will therefore be

$$y = k_2 \cdot x_2 = k_2 \cdot (x \pm x_1) = k_2 \cdot (x \pm k_1 \cdot y) \tag{1.15}$$

$$y = k_2 \cdot x \pm k_2 \cdot k_1 \cdot y \tag{1.16}$$

$$y \mp k_1 \cdot k_2 \cdot y = k_2 \cdot x \tag{1.17}$$

$$y(1 \mp k_1 \cdot k_2) = k_2 \cdot x \tag{1.18}$$

We obtain for positive feedback

$$y = \frac{k_2 \cdot x}{1 - k_1 \cdot k_2} \tag{1.19}$$

and for negative feedback

$$y = \frac{k_2 \cdot x}{1 + k_1 \cdot k_2} \tag{1.20}$$

A graphic solution, shown in figure 1.8, can simply be obtained in a common chart. For known dependency $y = f(x_2)$, the value of y is found, substituted to dependency $x_1 = f(y)$, and x_1 is added or subtracted to x (dependent on positive or negative feedback).

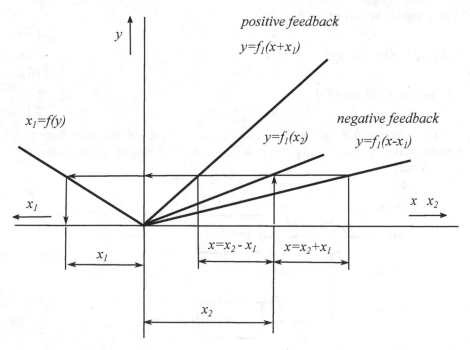

FIGURE 1.8: Graphic solution for feedback connection

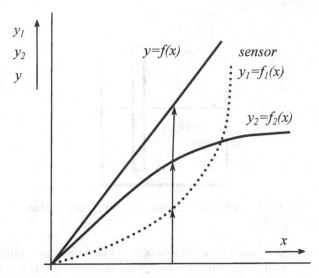

FIGURE 1.9: Block algebra used to linearize a non-linear steady-state characteristic

1.1.3 An example of block algebra

One of the examples showing how block algebra can be used is to linearize non-linear steady-state characteristics. This is shown in figure 1.9.

The sensor has a non-linear steady-state characteristic $y_1 = f_1(x)$. The goal is to linearize it to $y = f(x)$. The parallel connection will be used in the shown example, but other types can be used as well.

A block with a steady-state characteristic $y_2 = f_2(x)$ will be used. Summing y_1 and y_2 yields $y = f(x)$.

In some cases, the steady-state characteristic $y_2 = f_2(x)$ may be hard to create with simple devices. Then it can be created, for example, as a table in a micro-controller.

1.2 Static properties

1.2.1 Steady state

Let's imagine the following experiment, shown in figure 1.10. The temperature is T_1 and is constant. The water temperature in a tank is measured with a temperature sensor; the input signal will be denoted as X; the output signal will be denoted as Y.

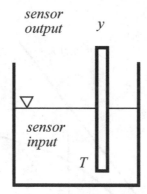

FIGURE 1.10: Experiment to define steady state

We begin the experiment with the sensor outside of the tank. A different temperature from T_1 will be measured. The actual temperature is not important. Now the sensor will be placed in the tank. It is obvious that the sensor will not immediately show the correct temperature. The sensor is placed in a thermowell. It does not have zero thermal capacity; the measured signal will slowly approach the correct temperature. In other words, there will be some delay before temperature T_1 will be measured correctly. This is called a transient response. Since this is an important topic, it will be handled in a separate chapter.

Sensor properties can be described in transients (functions of time) or in steady state (not functions of time). In the described experiment, we will wait until the input and output signals are steady. This will be called steady state.

Definition: Steady state is the state in which BOTH input signal AND output signal are constant, meaning they are not changing.

Now the question that anyone will ask is how long we need to wait before steady state is reached. As will be shown in the chapter about transients, this will depend on a parameter called "time constant." In theory it will be necessary to wait to infinity before the new steady state is reached. In practice, we will have to wait for five time constants to be in the new steady state with a sufficient accuracy.

The description of systems in steady state will allow us to use a simple description with algebraic equations. If the system will not be in steady state, differential equations will have to be used.

In the shown example, the new steady state would be reached when both temperature $T_1 = X$ and sensor output signal Y would be constant.

1.2.2   Steady-state characteristic

The system behavior in steady state will be described with a steady-state characteristic. It is the dependence of sensor output signal Y on sensor input

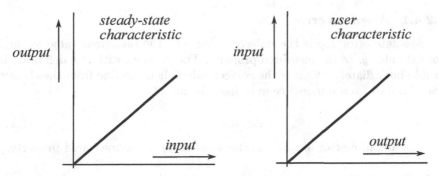

FIGURE 1.11: Example of a steady-state and user characteristic

signal X, $Y = f(X)$ in steady state. In the previously used example, it would be, for example, electric resistance = f(temperature T_1). An example of a steady-state characteristic is shown in figure 1.11.

The steady-state characteristic can also be obtained with an experiment. It can also be available as a chart or table, not a mathematical function. This is common in datasheets for sensors. The steady state characteristic can also be used in the block algebra.

1.2.3 User characteristic vs. steady-state (static) characteristic

Both characteristics are in steady state, but the static characteristic shows the dependency of sensor output on sensor input. In the case of our water temperature sensor the input is temperature. For a resistive temperature detector the output is electrical resistance.

Sometimes the user is more interested in an inverse dependency. We want to know what is the water temperature if we have measured resistance 100 Ω on a Pt100 resistive temperature sensor. This is the user characteristic. It is a function describing the relation of sensor input on sensor output.

An example of the user characteristic is shown in figure 1.11. The user characteristic can also be defined in many ways: mathematically, table, chart, etc.

Remember! Both are **IN STEADY STATE!**

1.2.4 Accuracy

Each measurement should have an estimate of its accuracy. Two approaches will be shown. The historically older one uses absolute and relative errors; the newer one uses uncertainties.

1.2.4.1   Absolute error

Absolute error Δy is the difference between the measured value y_m and correct value y_c of the measured property. The problem with this definition is visible immediately. What is the correct value? Is it the one from the theory, one obtained from a more accurate instrument,...?

$$\Delta y = y_m - y_c \qquad (1.21)$$

The absolute error always has the same units as the measured property.

1.2.4.2   Relative error

Relative error $\delta(y)$ is the ratio between absolute error Δy to the instantaneously measured value y_m. It is always in percent.

$$\delta(y) = \frac{\Delta y}{y_m} \cdot 100\,[\%] \qquad (1.22)$$

The definition has the same problem as absolute error. We don't know what the correct value is.

1.2.5   Instrument accuracy

1.2.5.1   Analog instruments—accuracy class

Accuracy class T_p expresses the maximal relative error of a value measured with a full deflection of the instrument's needle on a given range. It can be used to calculate maximal absolute error and other parameters of the instrument. It is expressed in percent. It is expressed as a number rounded to a number from a numerical series, typically 0,1; 0,5; 1; 2; 5 (percent).

The accuracy class T_p is calculated from the absolute error Δy_{max} and from range $y_{max} - y_{min}$

$$T_p = \frac{\Delta y_{\max}}{y_{\max} - y_{\min}} \cdot 100\,[\%] \qquad (1.23)$$

The accuracy class is typically shown directly on the instrument's scale, as shown in figure 1.12. It shows an analog voltmeter with $T_p = 0,5$.

The accuracy class allows us to calculate both absolute and relative errors.

Calculation example: an analog voltmeter with accuracy class 1 is used with range 300 V. The measured voltage is 60 V. How large is the absolute and relative error?

Absolute error $\Delta y_{\max} = T_p \cdot (y_{\max} - y_{\min})/100 = 1 \cdot (300 - 0)\,/100 = 3[V]$

Relative error $\delta\,(y) = \frac{\Delta y_{\max}}{x} \cdot 100\,[\%] = \frac{3}{60} \cdot 100\,[\%] = 5[\%]$

Relative error is 5%. It is significantly higher than would be directly obvious from the accuracy class. The reason for this is that we measure on the lower part of the scale, far away from the maximal deflection. The relative error is getting smaller as we approach the end of the scale. For this reason, **with an**

FIGURE 1.12: Analog voltmeter with accuracy class 0.5, 1 for range 600 V

analog instrument always try to measure with a maximal deflection of the needle (match the instrument range to the measured value).

1.2.5.2   Digital instruments

Digital instruments make use of an analog/digital (AD) converter. The converter converts the input voltage or current to a number that is shown on the display. The accuracy depends in majority on the A/D converter resolution.

Accuracy for a digital instrument is given by:
- accuracy of the analog to digital converter, e.g., 0,1% (given by the number of bits of the converter)
- error caused by the number of least significant digits of the device, i.e., 2 digits

All these parameters have to be found in the instrument manual. Also information about accuracy calculation has to be found. Some manufacturers give the accuracy as a), others as b):

a) $\pm$ (x % of reading + n digits)
b) $\pm$ (x % of range + n digits)

In some cases a combination of both is used, e.g., $\pm$ (x % of range + % y of reading + n digits)

TABLE 1.1: Example accuracy for a digital voltmeter

range	resolution	accuracy
600 mV	0,1 mV	(0,5% of reading + 8 digits)
6 V	1 mV	(0,8% of reading + 5 digits)
60 V	10 mV	(0,8% of reading + 5 digits)
600 V	100 mV	(0,8% of reading + 5 digits)
1000 V	1 V	(1% of reading + 10 digits)

Digit is the smallest part the instrument can measure. It is related to the A/D converter accuracy.

An example for a digital voltmeter is shown in table 1.1.

Example: The selected digital voltmeter has range 60 V, resolution 10 mV and accuracy ± (0,8% of reading + 5 digits). The measured value is 55,3 V. Calculate absolute error.

Answer: ± (0,8% of reading + 5 digits) = ± (0,008 · 55,3 + 5 · 0,01) = ±0,49 V

1.2.6   Sensitivity

Sensitivity is the capability of a system (sensor, instrument) to change its output for a change of input. It can be obtained from a steady-state characteristic. It has a physical meaning, like for example, how many Ω of change will I get if the temperature will change by 1°C (example of a resistance temperature sensor)?

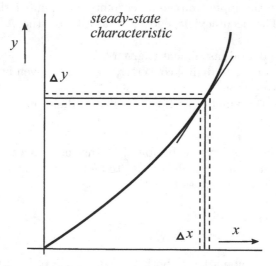

FIGURE 1.13: Sensitivity calculation

The sensitivity can be derived from the steady-state characteristic as shown in figure 1.13.

In a general case, sensitivity is calculated as

$$s = \lim_{\Delta x \to 0} \frac{\Delta y}{\Delta x} \qquad (1.24)$$

The units of sensitivity are the units of the output to the units of the input, i.e. $\Delta y / \Delta x$. For example, in case of a Pt100 resistive temperature sensor, the sensitivity is 0,385 $\Omega/$ °C.

For linear steady-state characteristics, the sensitivity is constant; for a non-linear steady state characteristic it is not constant.

1.3 Dynamic properties

1.3.1 Transient responses

Definition: the transient response is the response of a system to a step change on its input

Imagine the following experiment shown in figure 1.14. We have two tanks with water, one with temperature T_1, the second with temperature T_2. A temperature sensor is placed in a tank with cold water. The temperature T_1 in this tank is known and steady. We move the sensor to a second tank with

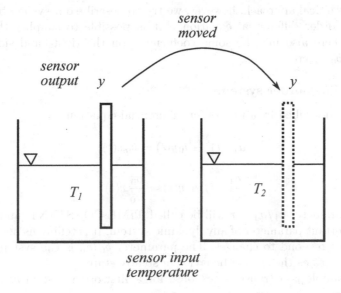

FIGURE 1.14: Example experiment to get the transient response

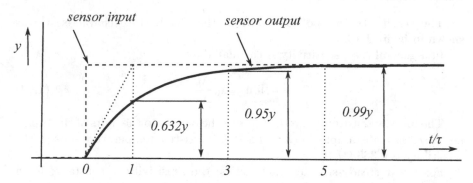

FIGURE 1.15: Transient response of a 1th order system

hot water, also with a known and steady temperature T_2. If this transition is fast, we can say that the sensor input signal has changed with a step. The output signal, however, will not be a step but a transient function.

If the sensor output y as a function of time is plotted, a transient response is visible, as shown in figure 1.15. Based on the system properties it will take a shorter or longer time before a new steady state is reached. No matter how good or fast the system is, there will always be a transient response, and it will always take some time before a new steady state is reached.

In general the transient response can be described by a n-th order differential equation

$$a_n y^{(n)} + \ldots + a_2 y'' + a_1 y' + a_0 y = x \tag{1.25}$$

In a practical approach, however, we try to describe our system by a first- or second-order differential equation if it is possible to simplify the calculations. There also may be some coefficients on the right-hand side of the equation b_0, b_1 etc.

1.3.1.1 First-order systems

This is described by a first-order differential equation.

$$a_1 y(t)' + a_0 y(t) = b_0 x(t) \tag{1.26}$$

$$\frac{a_1}{a_0} y(t)' + y(t) = \frac{b_0}{a_0} x(t) \tag{1.27}$$

The parameter $a_1/a_0 = \tau$ will be called TIME CONSTANT, and it is the most important parameter of any dynamic system. It is telling us how fast the system can respond to changes. The parameter b_0/a_0 is the system's static gain. It describes the system behavior in steady state.

An example is a transient response for a first-order system as shown in figure 1.15.

The y-axis in figure 1.15 is not shown as time but as a ratio of τ to time. This allows us to compare different systems. The input signal changes with a step change, and the output follows. In the shown example, the static gain is set to 1; in a general case, this may be different.

By solving the differential equation, we obtain the solution as an exponential function

$$y(t) = y_\infty \left(1 - e^{-t/\tau}\right) \tag{1.28}$$

where y_∞ is the output steady-state value.

It can be seen that the system's speed is given by a single parameter, the time constant τ. After one time constant has elapsed, the system is, however, just at $0{,}632\%$ of the new steady state.

$$1\tau =\sim 0,632\% y_\infty$$
$$3\tau =\sim 0,95\% y_\infty$$
$$5\tau =\sim 0,99\% y_\infty$$

From the differential equation it can also be seen that the new steady state is reached in infinite time. This is very impractical. However, it can also be seen that 99% of the new steady state is reached after more than 5 τ. Therefore, 5 τ is generally considered a new steady state with sufficient accuracy.

The time constant can also be found from experimental data. If we make a tangent at any point, the time it takes from this point until the time the tangent intersects the new steady-state value is the time constant. This is shown in figure 1.15. The tangent can be made at an arbitrary point.

1.3.1.2 Dynamic errors

The difference between instantaneous output signal y and output steady-state value y_∞ is called dynamic error

$$\Delta y_d(t) = y(t) - y_\infty = -y_\infty e^{-t/\tau} \tag{1.29}$$

It can be seen that the dynamic error is decreasing in time. It is highest for $t = 0$ and zero for steady state.

1.3.1.3 Second-order systems

More complex systems can be described by a second-order differential equation

$$a_2 y(t)'' + a_1 y(t)' + a_0 y(t) = b_0 x(t) \tag{1.30}$$

Based on the equation roots we can obtain different solutions.

a) Real different roots—over-damped system The solution is an aperiodic function.

$$\lambda_1 \neq \lambda_2 \tag{1.31}$$

$$y(t) = C_1 \cdot e^{\lambda_1 t} + C_2 \cdot e^{\lambda_2 t} + y_\infty \tag{1.32}$$

b) Real equal roots—critically damped system The solution is a function with minimal damping but without oscillations. It is the fastest achievable response without oscillations.

$$\lambda_1 = \lambda_2 \tag{1.33}$$

$$y(t) = (C_1 + C_2 \cdot t) \cdot e^{\lambda t} + y_\infty \tag{1.34}$$

c) Complex conjugate pole pair—under-damped system The response is oscillatory. This response is typically faster that aperiodic response but has some overshoot.

$$\lambda_{1,2} = \alpha \pm j\omega \tag{1.35}$$

$$y(t) = (A \cdot \cos \omega t + B \cdot \sin \omega t) \cdot e^{\alpha t} + y_\infty \tag{1.36}$$

A comparison for a second-order system is shown in figure 1.16.

By comparing the transient response for a first-order system in figure 1.15 and second-order system in 1.16, a difference is visible. The first-order system reacts immediately; the slope at starting time is non-zero. On the other hand, the slope for a second-order system at time 0 is zero. The system does not respond immediately.

1.3.2 Ramp responses

In some cases, it is not possible to obtain the system's step response, or it is better to change the system input signal not stepwise but linearly and to watch system output. We obtain also a transient response, but to a linear input signal. The ramp response is shown in figure 1.17. An example of this system is a heated tank with a temperature sensor in which the temperature is changed linearly.

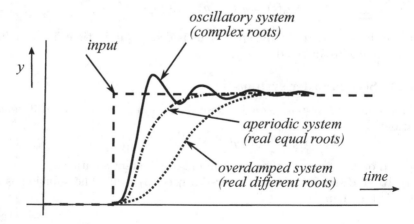

FIGURE 1.16: Transient response for a second-order system

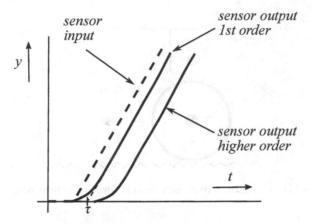

FIGURE 1.17: Ramp response

Sensor input is a linear function of time

$$x = w \cdot t \tag{1.37}$$

where w is a constant describing the slope.
For a first-order system, the solution of a differential equation describing the system is

$$y(t) = w \cdot \left[1 - \tau \cdot \left(1 - e^{-t/\tau}\right)\right] \tag{1.38}$$

Dynamic error is

$$\Delta y_d(t) = -w \cdot \tau \cdot \left(1 - e^{-t/\tau}\right) \tag{1.39}$$

and is minimal for $t = 0$ (start of transient). The maximal dynamic error is $\Delta y_{dmax} = -w \cdot \tau$ for $t - > \infty$

After the transient fades, the output follows the input with a constant difference. The difference between a first-order system and second-order system is also visible. A first-order system reacts immediately (non-zero slope of the tangent), the second-order system has initially a zero slope of the tangent.

1.3.3 Frequency characteristics

Frequency characteristics describe the system's response in amplitude and frequency for a periodic input signal of a fixed amplitude and phase with variable frequency. Imagine a system composed of a wheel, spring and damper, shown in figure 1.18. This can be a car on a road. The input signal x will be the uneven road surface, the output signal y is the position of the car's cabin floor. The input signal can be described as a harmonic function (or a sum of harmonic functions).

$$\bar{x} = x_0 \sin \omega t = x_0 e^{j\omega t} \tag{1.40}$$

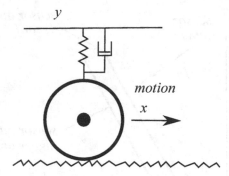

FIGURE 1.18: Frequency characteristic experiment

In linear systems, the output signal is also a harmonic function with the same frequency but a different amplitude and phase.

$$\bar{y} = y_0 \sin(\omega t + \varphi) = y_0 e^{j(\omega t + \varphi)} \tag{1.41}$$

The relation between output and input of a system can be described with transfer function. Transfer function mathematically describes relation of output signal to input signal.

$$G(j\omega) = \frac{\bar{y}}{\bar{x}} = \frac{y_0 \sin(\omega t + \varphi)}{x_0 \sin \omega t} = \frac{y_0 e^{j(\omega t + \varphi)}}{x_0 e^{j\omega t}} = A \cdot e^{j\varphi} \tag{1.42}$$

1.3.3.1 First-order systems

The system can also be described with a graph, the frequency characteristic in complex plane.

$$G(j\omega) = \frac{k}{1 + j\omega\tau} \tag{1.43}$$

The frequency characteristic for a first-order system is shown in figure 1.19. Frequency zero (DC signal) is on the real axis. With the increasing frequency the vector follows a circle in the complex plane. The point $\omega = 1/\tau$ is shown.

It shows the relation between the transient and frequency characteristics. A slow system (with a large time constant) will not be able to react to fast changes; it will not be able to follow high-frequency signals.

For frequency $\omega \to \infty$, the point is at the origin. The system can't follow an infinitely high frequency.

1.3.3.2 Second-order system

There are three possible solutions based on roots of the differential equation. The frequency characteristic in a complex plane is shown in figure 1.20.

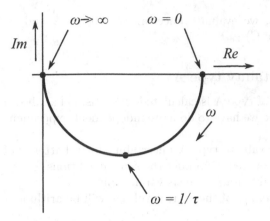

FIGURE 1.19: Frequency characteristic for a first-order system in a complex plane

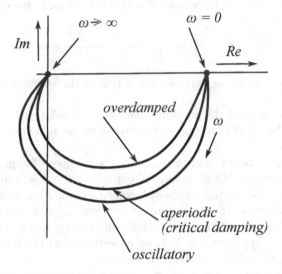

FIGURE 1.20: Frequency characteristic for a second-order system in a complex plane

1.4 Uncertainty of measurements

For every precise measurement, we need to know with what accuracy we have obtained our result. Uncertainty of measurement does not only depend on accuracy of used instruments, but also on the selected experimental method and on random influences that cannot be anticipated. According to evaluation of uncertainties, we evaluate uncertainty of type A and B (and combined uncertainty type C).

1.4.1 Uncertainty type A

Uncertainty of type A is calculated by statistical analysis of the measured values. Therefore we have to do more independent experiments with the same precision.

Example: Evaluate type-A uncertainty for a battery voltage. The measurement was repeated 10x under the same conditions.

The measured voltages are shown in table 1.2.

The best estimate of the correct voltage will be arithmetic mean

$$\overline{x} = \frac{\sum_{i=1}^{n} x_i}{n} = \frac{13,6}{9} = 1,51\,\mathrm{V} \tag{1.44}$$

The estimated standard deviation of arithmetic mean = uncertainty type A

$$u_A = s_{\overline{x}} = \frac{s_x}{\sqrt{n}} = \sqrt{\frac{\sum_{i=1}^{n}(x_i - \overline{x})^2}{n \cdot (n-1)}} = 0,3\,\mathrm{V} \tag{1.45}$$

The standard deviation determines the width of the interval where the data were.

Our best estimate of the measured voltage is therefore 1.51 V, with uncertainty 0.3 V. The results are typically rounded to the same number of decimal places.

In type-A uncertainty calculations also, the number of experiments has to be taken into account. If the number of experiments is low, our confidence in the result will be low. If the number of experiments is high, our confidence in the result will increase. If our experiment was done let's say 3x, our confidence will definitely be lower than if the experiment was repeated 100x. Therefore, if the number of experiments is low, we have to extend the interval.

The used approach follows the document [1].

Our experiment was repeated 9x, to achieve the desired confidence level the interval has to be extended.

The corrected type A uncertainty is

$$u_{U1Ak} = k^* \cdot u_{1A} \tag{1.46}$$

The coverage factor k is found by calculating degrees of freedom [1].

TABLE 1.2: Measured voltages

i	1	2	3	4	5	6	7	8	9	$\sum$
V (V)	1,49	1,53	1,49	1,50	1,49	1,56	1,55	1,50	1,49	13,6
$(x_i - \overline{x})^2$	0,78	0,14	0,78	0,32	0,78	1,76	1,02	0,32	0,78	6,68

TABLE 1.3: Value of $t_p(v)$ from the t-distribution for degrees of freedom v that defines an interval $-t_p(v)$ to $+t_p(v)$ that encompasses the fraction p of the distribution—from [1]

Degrees of freedom v	Fraction p [%]					
	68.27[1]	90	95	95.45	99	99.73
1	1.84	6.31	12.71	13.97	63.66	235.80
2	1.32	2.92	4.3	4.53	9.92	19.21
3	1.20	2.35	3.18	3.31	5.84	9.22
4	1.14	2.13	2.78	2.87	4.60	6.62
5	1.11	2.02	2.57	2.65	4.03	5.51
6	1.09	1.94	2.45	2.52	3.71	4.90
7	1.08	1.89	2.36	2.43	3.50	4.53
8	1.07	1.86	2.31	2.37	3.36	4.28
9	1.06	1.83	2.26	2.32	3.25	4.09
10	1.05	1.81	2.23	2.28	3.17	3.96
25	1.02	1.71	2.06	2.11	2.79	3.33
30	1.02	1.70	2.04	2.09	2.75	3.27
35	1.01	2.03	2.07	2.72	2.72	3.23
40	1.01	1.68	2.02	2.06	2.70	3.20
45	1.01	1.68	2.01	2.06	2.69	3.18
50	1.01	1.68	2.01	2.05	2.68	3.16
100	1.005	1.660	1.984	2.025	2.626	3.077
∞	1.000	1.645	1.960	2.000	2.576	3.000

degrees of freedom v

$$v = n - 1 \qquad (1.47)$$

where n is the number of experiments

The shown approach assumes Student t-probability distribution and is according to [1].

From table 1.3 the coverage factor for the calculated degrees of freedom and the desired confidence level is found.

The given example is for the result in format "best estimate"; "uncertainty."

If the result should be given in format "best estimate" $\pm$ "uncertainty," then the values in the table have to be divided by 2. This assumes a symmetric interval.

1.4.2 Uncertainty type B

The uncertainty of type B is used for non-statistical processing of measured data. There are several possibilities of calculations. It can be based on rational judgments from the properties of measuring instruments, experience from previous measurements, data obtained in calibration certificates, data sheets, reference manuals, etc. In this case we directly estimate the sizes of individual uncertainties and then with the law of propagation of uncertainty calculate total uncertainty type B.

In some cases the uncertainty is known from the instrument's manufacturer, datasheet, calibration certificate, etc. Uncertainty can also be estimated from accuracy (with certain assumptions).

How to calculate uncertainty from accuracy If it is possible to estimate only the interval where the measured value "has to be" with certainty "almost 100%" then standard uncertainty type B u_B can be calculated from the estimation of type of probability distribution function (normal, rectangular, triangular, etc.) and from known deviations Δ from the mean value.

$$u = \frac{\Delta}{k} \tag{1.48}$$

where Δ is the half width of the interval ((upper limit – lower limit) / 2), and k is the coverage factor describing the selected probability distribution.

Some examples of those coefficients for selected probability distributions are in table 1.4.

The normal distribution is a function that represents the distribution of many random variables as a symmetrical bell-shaped graph where the peak is centered about the mean and is symmetrically distributed in accordance with the standard deviation [2].

The rectangular distribution is a function that represents a continuous uniform distribution and constant probability. In a rectangular distribution, all outcomes are equally likely to occur [2].

The U-shaped distribution is a function that represents outcomes that are most likely to occur at the extremes of the range [2].

The triangle distribution is a function that represents a known minimum, maximum and estimated central value. It is commonly referred to as the "lack of knowledge" distribution because it is typically used where a relationship between variables is known, but data is scare [2].

TABLE 1.4: Selected values for k for various probability distribution, [2]

probability distribution	uniform (for probability 57.74%)	normal (Gauss) for 2Σ (95.45%)	U-shape	triangular
k	$\sqrt{3}$	2	$\sqrt{2}$	$\sqrt{6}$

Example: Calculation of type B uncertainty for a power supply voltage

The power supply specifications define for the unloaded output the voltage 9.15 V ± 0.15 V. The uncertainty is not given, but the interval is. Therefore the uncertainty can be calculated.

The given interval 9,15 V ± 0,15 V is the interval where the voltage will be with almost 100% confidence. The voltage is an analog property, therefore the normal (Gaussian) distribution will be used. For confidence interval 2σ, coverage factor is $k = 2$, and uncertainty is

$$u_U = \frac{\Delta}{k} = \frac{0.15}{2} = 0.075V \tag{1.49}$$

Example: Calculation of type B uncertainty for a digital voltmeter

The specification of a digital voltmeter states for range 400 mV the resolution 100 μV (digit). Accuracy is ±(0,5% of reading + 4 digits). The measured voltage is 88.2 mV.

Absolute error is

$$\Delta U = \frac{0,5 \cdot of\,reading}{100} + 4 \cdot digits = \frac{0.5 \cdot 88.2mV}{100} + 4 \cdot 100\mu V = 841\mu V \tag{1.50}$$

For a digital voltmeter the uniform probability distribution will be used (same probability of error in the whole interval). The uncertainty is

$$u_B = \frac{\Delta U}{\sqrt{3}} = \frac{841\mu V}{\sqrt{3}} = 486\mu V \tag{1.51}$$

The calculation is done for confidence level 57.74%.

Example: Calculation of type-B uncertainty for a resistor A resistor has a nominal value $R = 270$ kΩ, tolerance ± 0.1%. Absolute error is

$$\Delta_R = R \cdot 0,1\%/100 = \pm 0,27k\Omega \tag{1.52}$$

Assuming normal (Gaussian) distribution with 2σ, the uncertainty is

$$u_R = \frac{\Delta_R}{k} = \frac{0,27}{2} = 0,135k\Omega \tag{1.53}$$

1.4.3 Law of propagation of uncertainty

In a general case, when the calculated variable depends on input variables $x_1, x_2, ..., x_n$, the law of propagation of uncertainty is used.

$$u = \sqrt{\left(\frac{\partial f}{\partial x_1} \cdot u_{x_1 N}\right)^2 + \left(\frac{\partial f}{\partial x_2} \cdot u_{x_2 N}\right)^2 + ... + \left(\frac{\partial f}{\partial x_n} \cdot u_{x_n N}\right)^2} \tag{1.54}$$

where $u_{X_i N}$ is the uncertainty of the nominal value of the variable x_n.

If one uncertainty is significantly larger than the others, the other variables will not have a large effect. The largest uncertainty will be dominant. No improvement in the small uncertainties will improve the final uncertainty significantly. The dominant uncertainty has to be decreased first to achieve an improvement.

Equation 1.54 is valid only when variables $x_1, x_2, ..., x_n$ are not correlated. If individual variables are correlated, covariances have to be calculated and the law takes a more general form

$$u_B = \sqrt{\sum_{i=1}^{m} A_i^2 u^2(x_i) + 2\sum_{i=2}^{m}\sum_{j<i}^{m-1} A_i A_j u(x_i, x_j)} \qquad (1.55)$$

where $u(x_i, x_j)$ are covariance coefficients between variables x_i and x_j, and A_i are sensitivities

$$A_i = \frac{\partial(X_1, X_2, ..., X_m)}{\partial X_i}\bigg|_{X_1=x_1,...,X_m=x_m} \qquad (1.56)$$

Example: Uncertainty calculation for a formula with two variables

For circuit in figure 1.21 calculate type B uncertainty for resistance R_2. Power supply voltage is 9.15 V, its uncertainty is 0,075 V. The voltage U1 was measured 86,2 mV, corrected type-A uncertainty is 0,2 mV. Resistance R1 = 270 kΩ, uncertainty 0,135 kΩ.

Circuit analysis yields for R_2

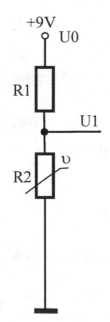

+9V
U0

R1

U1

R2

FIGURE 1.21:
Voltage divider

$$R2 = \frac{R1 \cdot U1}{U0 - U1} \qquad (1.57)$$

Partial derivatives are

$$\frac{\partial R2}{\partial R1} = \frac{U1}{U0 - U1} \qquad (1.58)$$

$$\frac{\partial R2}{\partial U1} = \frac{U0 \cdot R1}{(U0 - U1)^2} \qquad (1.59)$$

$$\frac{\partial R2}{\partial U0} = \frac{-U1 \cdot R1}{(U0 - U1)^2} \qquad (1.60)$$

The partial derivatives will be used to calculate the contributions of individual components to the result.

$$u_{R2} = \sqrt{\left(\frac{\partial R2}{\partial R1} \cdot u_{R1}\right)^2 + \left(\frac{\partial R2}{\partial U1} \cdot u_{U1}\right)^2 + \left(\frac{\partial R2}{\partial U0} \cdot u_{U0}\right)^2} \qquad (1.61)$$

For U0, R1 nominal values will be substituted (9.15 V and 270 kΩ, respectively). U1 = 86,2 mV. For u_{R1}, u_{U0}, u_{U1} the known uncertainties are substituted.

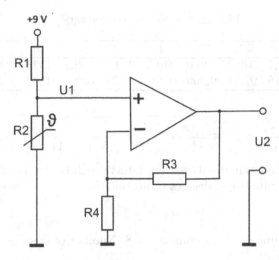

FIGURE 1.22: Used schematic for uncertainty calculation

1.4.4 Example for a more complicated connection

The calculation for a more complicated connection will be shown for the connection in figure 1.22. The resistor R2 is a temperature sensor, thermistor type B57164K [3] with nominal resistance for 25°C $R_{25} = 2200\Omega \pm 10\%$. The thermistor has a non-linear steady-state characteristic, for the used type it is No. 1013 in the datasheet [3].

The connection is powered from a power supply with nominal voltage 9V and power supply type RQT666K. The operation amplifier used is TS27L2CN [4]. The voltage $U2$ is measured with a voltmeter type MT-1232 [5] and used range 400 mV DC.

All the used resistors are precision metal film resistors with tolerances $\pm$ 0,1%. Nominal resistor values are $R1 = 270$ kΩ, $R3 = 1,2$ kΩ, $R4 = 1,2$ kΩ.

The goal is to calculate type-A uncertainty from the measured voltage U_2 and type-B uncertainty from the following sources:
- uncertainty caused by the power supply voltage U_0
- uncertainty caused by R1, R2 tolerance
- uncertainty caused by R3, R4 tolerance (consider the operational amplifier as ideal)

The calculation should be done for a constant temperature 20°C.
Uncertainty type-A calculation
The measured voltages U_2 are in table 1.5.
Arithmetic mean = best estimate of the correct value

$$\overline{x} = \frac{\sum_{i=1}^{n} x_i}{n} = \frac{2716}{8} = 339,5 \text{ mV} \tag{1.62}$$

TABLE 1.5: Measured voltages

i	1	2	3	4	5	6	7	8	$\sum$
U_2 (mV)	339,1	340,2	340,0	340,1	339,3	339,2	339,2	338,9	2716
$(x_i - \bar{x})^2$	0,119	0,571	0,309	0,430	0,021	0,060	0,296	0,554	2,36

$$u_A = s_{\bar{x}} = \frac{s_x}{\sqrt{n}} = \sqrt{\frac{\sum_{i=1}^{n}(x_i - \bar{x})^2}{n \cdot (n-1)}} = \sqrt{\frac{2,36}{8 \cdot (8-1)}} = 0,2 \text{ mV} \qquad (1.63)$$

The voltage was measured only 8x; the interval will be extended.
Based on [1] the number of degrees of freedom is

$$\nu = n - 1 \qquad (1.64)$$

where n is the number of experiments = 8. Number of degrees of freedom is therefore $\nu = 7$

For the selected probability distribution (normal = Gaussian) and for the selected confidence level (95,45%) the value is found in table G.2 in [1] => k = 2,43

$$u_A k = k \cdot u_A = 2,43 \cdot 0,2 \text{ mV} = 0,5 \text{ mV} \qquad (1.65)$$

Type-A uncertainty is:
- best estimate of the correct value 339,5 mV
- type-A uncertainty 0.5 mV (rounded to same number of decimal places as the best estimate).

The calculation was done assuming normal probability distribution with confidence level 95.45%.

Uncertainty type-B calculation
- Uncertainty caused by variation of the power supply voltage U_0
The power supply voltage is specified as 9,15 V $\pm$ 0,15 V when the power supply is unloaded. The given interval is where the voltage will be with "almost 100%" confidence. Assuming the normal probability distribution with 2σ, the coverage factor is $k = 2$ and uncertainty is

$$u_U = \frac{\Delta}{k} = \frac{0,15}{2} = 0,075 \text{ V} \qquad (1.66)$$

- Uncertainty caused by resistor R1, R2 tolerance
The nominal resistance for R1 = 270 kΩ, tolerance $\pm$ 0.1%. Absolute error is

$$\Delta_{R1} = R1 \cdot 0,1\%/100 = \pm 0,27 \text{ k}\Omega \qquad (1.67)$$

Assuming normal distribution with 2σ the uncertainty is

$$u_{R1} = \frac{\Delta_{R1}}{k} = \frac{0,27}{2} = 0,135 \text{ k}\Omega \qquad (1.68)$$

For R2 the calculation will be done for temperature 20°C.

The nominal resistance R2 = 2200 Ω, tolerance ± 10% is given for temperature 25°C. The resistance for 20°C is calculated using the equations from [3].

$R_T/R_{25} = 1,2403 => R_T(20°C) = 1,2403 \cdot R_{25} = 1,2403 \cdot 2200 = 2729Ω$

Tolerance $R2_{25°C}$ is $\Delta R2_{25°C} = R2 \cdot 10\%/100 = \pm272,9Ω$

Assuming normal probability distribution with 2σ the uncertainty is

$$u_{R2} = \frac{\Delta R2_{25°C}}{k} = \frac{272,9}{2} = 136 \ Ω \tag{1.69}$$

The voltage and its uncertainty on the voltage divider is now calculated. This will simplify the following calculations. Since an ideal op-amp is considered, its input resistance is infinite and hence the voltage divider is unloaded. With this assumption the voltage divider output voltage is

$$U1 = U0 \cdot \frac{R2}{R1 + R2} = 9,15 \cdot \frac{2729}{2729 + 270000} = 91,6 \ mV \tag{1.70}$$

The law of propagation of uncertainty 1.54 will now be used
In our case

$$u_B = \sqrt{(\frac{\partial U1}{\partial U0} \cdot u_{U1})^2 + (\frac{\partial U1}{\partial R1} \cdot u_{R1})^2 + (\frac{\partial U1}{\partial R2} \cdot u_{R2})^2} \tag{1.71}$$

Partial derivatives of equation 1.70 for variables $U1, R1$ and $R2$ are

$$\frac{\partial U1}{\partial U0} = \frac{R2}{R2 + R1} = \frac{2729}{2729 + 270000} = 0,01 \tag{1.72}$$

$$\frac{\partial U1}{\partial R1} = \frac{-R2 \cdot U0}{(R2 + R1)^2} = \frac{-2729 \cdot 9,15}{(2729 + 270000)^2} = -0,34 \cdot 10^{-6} \tag{1.73}$$

$$\frac{\partial U1}{\partial R2} = \frac{U0}{R2 + R1} - \frac{R2 \cdot U0}{(R2 + R1)^2} =$$
$$= \frac{9,15}{2729 + 270000} - \frac{2729 \cdot 9,15}{(2729 + 270000)^2} =$$
$$= 33,3 \cdot 10^{-6} \tag{1.74}$$

Type B uncertainty of the voltage U1 is

$$u_{U1_B} = \sqrt{(0,01 \cdot 0,075)^2 + (-0,34 \cdot 10^{-6} \cdot 136)^2 + (33,3 \cdot 10^{-6} \cdot 136)^2} =$$
$$= 4,6 \ mV \tag{1.75}$$

- Uncertainty caused by tolerance of R3, R4
Uncertainty R3 and R4 from tolerance

$$\Delta_{R3} = R3 \cdot 0,1\%/100 = \pm1,2Ω \tag{1.76}$$

$$\Delta_{R4} = R4 \cdot 0,1\%/100 = \pm1,2Ω \tag{1.77}$$

Assuming normal probability distribution with 2σ the uncertainty of R3 and R4 is

$$u_{R3} = \frac{\Delta_{R3}}{k} = \frac{1,2}{2} = 0,6\Omega \tag{1.78}$$

$$u_{R4} = \frac{\Delta_{R4}}{k} = \frac{1,2}{2} = 0,6\Omega \tag{1.79}$$

The amplifier output voltage is

$$U2 = (\frac{R3}{R4} + 1) \cdot U1 \tag{1.80}$$

Partial derivatives are

$$\frac{\partial U2}{\partial U1} = (\frac{R3}{R4} + 1) = \frac{1200}{1200} + 1 = 2 \tag{1.81}$$

$$\frac{\partial U2}{\partial R3} = \frac{U1}{R4} = \frac{0,0916}{1200} = 76,3 \cdot 10^{-6} \tag{1.82}$$

$$\frac{\partial U2}{\partial R4} = \frac{-U1 \cdot R3}{R4^2} = \frac{-0,0916 \cdot 1200}{1200^2} = 76,3 \cdot 10^{-6} \tag{1.83}$$

Therefore the uncertainty of the amplifier output voltage is

$$u_{U2} = \sqrt{(\frac{\partial U2}{\partial U1} \cdot u_{U1})^2 + (\frac{\partial U2}{\partial R3} \cdot u_{R3})^2(\frac{\partial U2}{\partial R4} \cdot u_{R4})^2} =$$
$$= \sqrt{(2 \cdot 4,6 \cdot 10^{-3})^2 + (76,3 \cdot 10^{-6} \cdot 0,6)^2 + (-76,3 \cdot 10^{-6} \cdot 0,6)^2} = \tag{1.84}$$
$$= 46,7 \text{ mV}$$

The amplifier output voltage is

$$U2 = (\frac{R3}{R4} + 1) \cdot U1 = (\frac{1200}{1200} + 1) \cdot 91,6 \text{ mV} = 183,2 \text{ mV} \tag{1.85}$$

2

Circuits for sensors

The sensor transfers the measured property (temperature, pressure, position, humidity, etc.) to some other physical property, mostly electrical (voltage, current, resistance, capacitance, inductance, etc.). This is shown in figure 2.1. From the point of view of any controller system, we like to have the sensor's output signal in some well-defined ranges. A standard signal is needed. Therefore electronic circuits are needed to convert the signal to some standard signal. The standard signal is then connected to a controller system. The use of standard signals allows interconnecting different sensors and controller systems together. The most common standard analog signals are summarized in table 2.1.

2.1 Voltage signals

The voltage signal V_1 is available at the sensor output. It is transferred to a voltmeter with input resistance R_L through connections with non-zero wire resistance R_w. There is a voltage drop dV on the wire resistances. This is shown in figure 2.2.

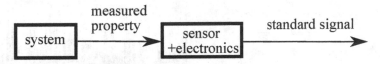

FIGURE 2.1: Importance of unified transducer output signal

TABLE 2.1: Standard analog signals—voltage and current
most common shown in bold

Voltage	Current
0–10 V	0–20 mA
−10–10 V	**4–20 mA**
0–5 V	
−5–5 V	

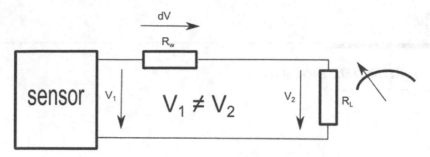

FIGURE 2.2: Sensor with voltage output

Therefore the voltmeter does not measure the same voltage as directly on the sensor output. Voltage drops limit the connection distance. The voltage signal is also sensitive to interference. It should therefore not be used in applications where the interference is large, such as near frequency inverters, motors, relays, etc. The interference can be partially limited with shielded cables. Still, the voltage signal is suitable only in applications where the interference level is small, such as labs, building air conditioning applications, etc. Industrial sensor installations use almost exclusively the current signals, which are significantly less sensitive to interference.

An example how the voltage signal is sensitive to interference is shown in figure 2.3. The left part of the figure shows a voltage signal, a thermocouple output. The thermocouple cable was placed in close proximity to a cable from a frequency inverter to a motor. The figure on the left was taken when the inverter was off. Figure 2.3 right shows the same signal, but when the frequency

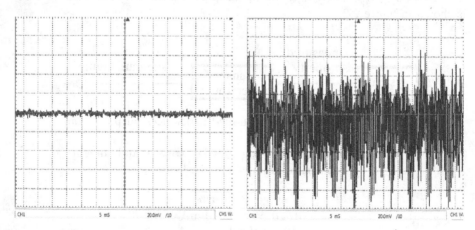

FIGURE 2.3: Example of voltage output, thermocouple output in this case—left = frequency inverter off, right = inverter on—signal unusable, signal distorted by interference from frequency inverter

inverter was on. The interference masks out all the signal from the sensor; it is unusable.

2.2 Current signals

The current signal is significantly less sensitive to interference than the voltage signal. Two ranges are used, 0 to 20 mA or 4 to 20 mA. The latter is more often. The advantage of the current signal is that following the Kirchhoff's law, the current is the same in all the connections. This is shown in figure 2.4.

The current in the loop is kept constant, it does not depend on the connecting wire resistance and therefore on the changes caused by temperature. There is however a limitation. The constant current is maintained by changing voltage. The constant current is typically maintained by an operational amplifier. The op-amp acts as a controller, it compares the set current with the measured current. When the wire resistance has changed, the op-amp compensates by adjusting its output voltage. Due to the limited power supply voltage and also for safety reasons, the voltage can't increase indefinitely. The output voltage is limited. Hence, also the maximal resistance in the current loop has to be limited. This will limit the maximal length of the current loop. The loop resistance is typically limited to 600 ohm (standard ANSI/ISA-50.00.01.1975 (R2002)). The receiver resistance is typically set to 100 ohm, leaving 500 ohm for the current loop. The length is therefore typically limited to a few hundred meters.

The sensor can also be powered directly from the current loop, assuming its power consumption is not larger than 4 mA. This obviously applies only to range 4 to 20 mA.

The 4 to 20 mA current loop also allows for a simple fault detection. When the wire is interrupted, the current is 0.

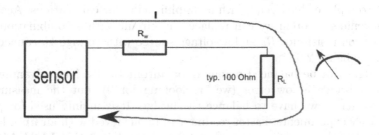

FIGURE 2.4: Current loop

For all those reasons, the 4 to 20 mA is the most common analog signal type used in industrial systems.

2.3 Bridge circuits

To process the sensor signal to a standard signal, electronic circuits are necessary. One of the most important circuits is a bridge circuit.

Bridge circuits can be divided into many categories. Based on the power supply voltage, there are DC, AC and impulse bridges. DC bridges are powered with DC voltage, AC bridges are powered with AC voltage and impulse bridges are powered with impulse voltage. Based on the output voltage detection, there are balanced and unbalanced bridges. In an unbalanced bridge, the output voltage is measured directly with a voltmeter. In a balanced bridge, the elements in a bridge are adjusted until zero voltage on the output is obtained.

2.3.1 DC bridges

DC bridges are used for sensors that change electrical resistance. Capacitive and inductive components are not present in DC circuits. The bridge acts as an amplifier, and transfers the small changes in resistance to a small voltage. The voltage typically needs to be amplified with a DC amplifier.

2.3.1.1 Wheatstone bridge

The bridge is formed by four resistances (one or more of those can be the sensors) and powered with a power supply V. Bridge output is V_0 and is usually connected to a voltmeter or some other processing circuit, such as a DC or instrumentation amplifier. Power supply resistance R_i is usually small compared to R_1 through R_4, so we can neglect it in a simplified circuit solution.

The Wheatstone bridge is shown in figure 2.5. The goal of the circuit analysis is to replace this circuit with a simplification for calculations. According to Thevenin's theorem we can replace any circuit with a combination of its substitution resistance R_k and a voltage source with voltage V_0 connected in series, see figure 2.6.

Bridges can be balanced manually or automatically. The advantage of a balanced bridge is low error (we are looking for 0), but the measurement takes longer as we have to balance the bridge. It is mainly used for precise laboratory experiments. Faster reading can be achieved with an **unbalanced bridge**, but the precision is lower and the **output is NON-LINEAR**.

Bridge output voltage—between nodes d and c is (for unloaded bridge = I output = 0)

$$V_0 = V_c - V_d \qquad (2.1)$$

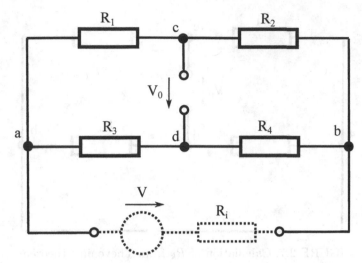

FIGURE 2.5: Wheatstone bridge

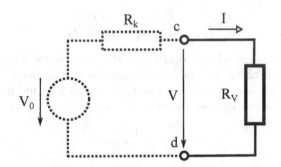

FIGURE 2.6: Bridge equivalent circuit diagram

$$V_c = \frac{R_1}{R_1 + R_2} \cdot V \tag{2.2}$$

$$V_d = \frac{R_3}{R_3 + R_4} \cdot V \tag{2.3}$$

By substituting we obtain

$$V_0 = \frac{R_1}{R_1 + R_2} \cdot V - \frac{R_3}{R_3 + R_4} \cdot V \tag{2.4}$$

To obtain the substitution resistance R_k, we have to short voltage sources and disconnect the current source according to Thevenin's theorem. The circuit in figure 2.7 is obtained.

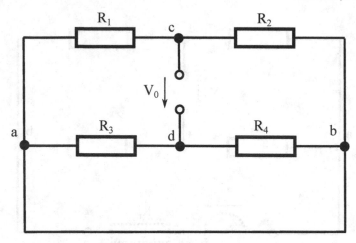

FIGURE 2.7: Calculation of R_k from Thevenin's theorem

Between terminal c and d we have resistance R_k

$$R_K = \frac{R_3 R_4}{R_3 + R_4} + \frac{R_1 R_2}{R_1 + R_2} \qquad (2.5)$$

The final equivalent circuit diagram is shown in figure 2.6.

The equivalent circuit diagram can be used to find properties of the circuit. The bridge output is connected to a voltmeter with internal resistance R_v.

The bridge output current therefore is

$$I = \frac{V_0}{R_K + R_v} \qquad (2.6)$$

$$I = V \frac{R_2 R_3 - R_1 R_4}{R_1 R_2 \left(R_3 + R_4 \right) + R_3 R_4 \left(R_1 + R_2 \right) + R_v \left(R_1 + R_2 \right) \left(R_3 + R_4 \right)} \qquad (2.7)$$

For a balanced bridge, output current I = 0. To achieve this, the following equation needs to be fulfilled. This is called equation of balance.

$$R_2 R_3 = R_1 R_4 \qquad (2.8)$$

For an unbalanced bridge, the output current is non-zero. The sensor has resistance R_0, variable by $\pm \Delta R$. **The dependence between output voltage and $\pm \Delta R$ is non-linear.**

2.3.1.2   Quarter-bridge

This connection uses one variable resistor (sensor). The other resistors are fixed (in the case of an unbalanced bridge), or at least one is variable (in the case of a balanced bridge).

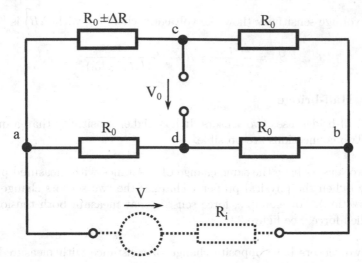

FIGURE 2.8: Quarter-bridge

Usually the following choice is made

$$R_1 = R_2 = R_3 = R_4 = R_0 \tag{2.9}$$

The sensor resistance is R_0, and it is variable by $\pm \Delta R$. 1/4 bridge is shown in figure 2.8.

The bridge output current is

$$I = V \frac{R_0 R_0 - (R_0 \pm \Delta R) R_0}{(R_0 \pm \Delta R) R_0 (R_0 + R_0) + R_0 R_0 ((R_0 \pm \Delta R) + R_0)} \tag{2.10}$$
$$+ R_v ((R_0 \pm \Delta R) + R_0)(R_0 + R_0)$$

It is obvious that the equation is non-linear for a linear change of ΔR. Moreover the measured voltage and current depends on the voltmeter internal resistance R_v.

For small changes of ΔR and assuming $\Delta R \ll R_0$ the equation can be simplified

$$I = V \frac{1}{4(R_v + R_0)} \cdot \frac{\pm \Delta R}{R_0} \tag{2.11}$$

The quarter-bridge current sensitivity (how the current is changing with ΔR) is

$$S_I = \frac{\Delta I}{\frac{\Delta R}{R_0}} = V \frac{1}{4(R_v + R_0)} \tag{2.12}$$

The quarter-bridge output voltage is

$$V_{CD} = R_V \cdot I = R_V \cdot V \frac{1}{4(R_v + R_0)} \cdot \frac{\pm \Delta R}{R_0} \tag{2.13}$$

and the voltage sensitivity (how the voltage is changing with ΔR) is [6]

$$S_V = \frac{\Delta V_{CD}}{\frac{\Delta R}{R_0}} = R_V \cdot V \frac{1}{4(R_v + R_0)} \tag{2.14}$$

2.3.1.3 Half-bridge

The half-bridge uses two sensors. It has higher sensitivity than a quarter-bridge. Two connections are possible.

a) the two sensors have the same change of resistance with measured physical property (when the physical property changes the two sensors change in the same direction). For example, force sensors that measure both tension with the applied force. See figure 2.9.

b) the two sensors have opposite change of resistance with measured physical property (when the physical property changes the two sensors change in the opposite direction). An example is force sensors when one sensor measures tension and the other one measures compression with an applied force. See figure 2.10. As will be obvious from the equations, this doubles the sensitivity.

Half-bridge current sensitivity is

$$S = \frac{\Delta I_v}{\frac{\Delta R}{R_0}} = V_0 \frac{2}{4(R_v + R_0)} \tag{2.15}$$

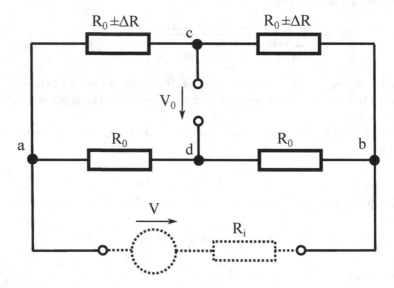

FIGURE 2.9: Half-bridge—two sensors measure in the same direction

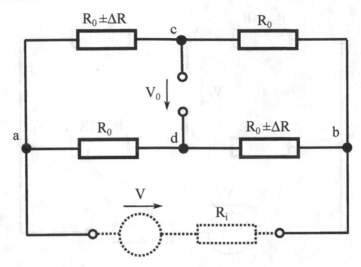

FIGURE 2.10: Half-bridge—two sensors measure in the opposite direction

Half-bridge voltage sensitivity is

$$S_V = \frac{\Delta V_{CD}}{\frac{\Delta R}{R_0}} = R_V \cdot V_0 \frac{2}{4\left(R_v + R_0\right)} \tag{2.16}$$

2.3.1.4 Full bridge

A full bridge will use four sensors. It is obvious that two sensors have to measure in one direction and two in the other direction with the same change of the input variable—see figure 2.11. For example, two strain gauges have to measure tension, and two have to measure compression caused by the same force. This has to be achieved with the correct placement of the sensors.

As all elements in the bridge are the sensors, there is no extra element for balancing. It is used as an unbalanced bridge.

Full-bridge current sensitivity is

$$S = \frac{\Delta I_v}{\frac{\Delta R}{R_0}} = V_0 \frac{4}{4\left(R_v + R_0\right)} \tag{2.17}$$

Full-bridge voltage sensitivity is

$$S_V = \frac{\Delta V_{CD}}{\frac{\Delta R}{R_0}} = R_V \cdot V_0 \frac{4}{4\left(R_v + R_0\right)} \tag{2.18}$$

The sensitivity doubles compared to the half-bridge and quadruples compared to the quarter-bridge.

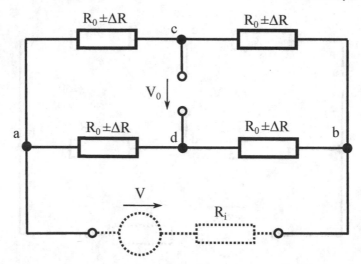

FIGURE 2.11: Full bridge

2.3.2   AC bridges

AC bridges use AC supply. They are used for AC variables such as inductance or capacitance. The bridge is supplied with voltage frequency in range 65 Hz to approximately 5 kHz. This limits the interference from the 50 or 60 Hz power supply network, lower frequencies are filtered out with a low-pass filter. An AC bridge is shown in figure 2.12.

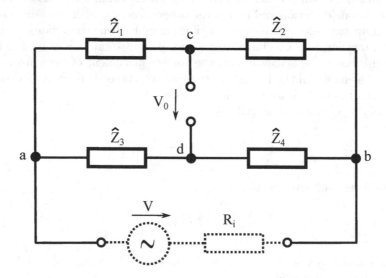

FIGURE 2.12: AC bridge

An AC bridge can be balanced or unbalanced. The condition of balance requires impedances to match. Amplitude and phase have to be matched. Therefore, an AC quarter bridge requires two elements for balancing.

$$\hat{Z}_1 \cdot \hat{Z}_4 = \hat{Z}_2 \cdot \hat{Z}_3 \tag{2.19}$$

$$|Z_1| \cdot |Z_4| = |Z_2| \cdot |Z_3| \tag{2.20}$$

$$\varphi_1 + \varphi_4 = \varphi_2 + \varphi_3 \tag{2.21}$$

AC bridge voltage sensitivity is

$$S = \frac{\Delta \hat{V}_{cd}}{\frac{\Delta \hat{Z}_1}{\hat{Z}_1}} \tag{2.22}$$

2.3.3 Differential bridges

A differential bridge is basically an AC bridge, but two elements in the bridge are replaced with two voltage sources. A differential bridge is shown in figure 2.13.

The condition of balance is

$$\hat{V}_1 \cdot \hat{Z}_2 = \hat{V}_2 \cdot \hat{Z}_1 \tag{2.23}$$

Assuming $\Delta \hat{Z} << \hat{Z}$, the sensitivity of the differential bridge is

$$S = \frac{\Delta I}{\frac{\Delta \hat{Z}}{\hat{Z}}} = \frac{V}{\hat{Z} + \Delta \hat{Z}} \tag{2.24}$$

One example of such a sensor is a linear variable differential transformer (LVDT). It is a position sensor composed of a transformer with one primary winding (AC powered), two secondary wingdings and a movable core. The secondary wingdings are connected into a differential bridge with two resistances. When the core is in a central position, same voltages are induced to

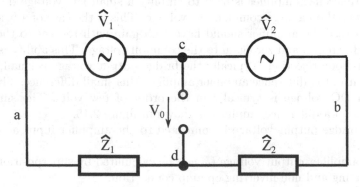

FIGURE 2.13: Differential bridge

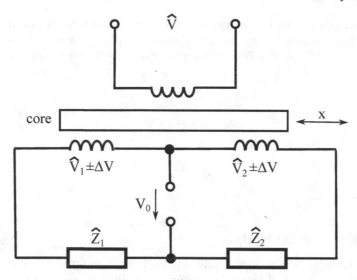

FIGURE 2.14: Differential bridge—LVDT

the secondary wingdings and the bridge is balanced. When the core moves in either direction, the voltages induced in the secondary wingdings differ and the bridge becomes unbalanced. The output voltage of the bridge is measured with a voltmeter. It is proportional to position of the core. The differential bridge—LVDT—is shown in figure 2.14.

2.4   Difference amplifier

The difference amplifier is used to amplify a small DC voltage difference superimposed on a large common DC voltage. This is the case of a bridge. In the ideal case the amplifier should have high gain with respect to the small voltage difference, and zero gain to the common voltage, This ability is called common-mode rejection. Typically the bridge output voltage is small, in the order of mV, the difference amplifier amplifies this small difference. The large common DC voltage is typically in the order of few volts. The simplified schematic of a difference amplifier is shown in figure 2.15.

The bridge output voltage is connected to the amplifier input as voltage u_{11}, u_{12}.

The amplifier output voltage U_2 can be calculated by superposition, from the inverting and non-inverting op-amp connection.

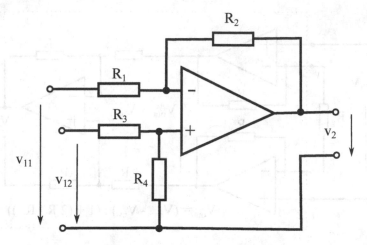

FIGURE 2.15: Difference amplifier

The amplifier output voltage is

$$u_{21} = \frac{-R_2}{R_1} u_{11} \tag{2.25}$$

$$u_{22} = \left(1 + \frac{R_2}{R_1}\right) \frac{R_4}{R_3 + R_4} u_{12} \tag{2.26}$$

$$u_2 = u_{21} + u_{22} = \left(1 + \frac{R_2}{R_1}\right) \frac{R_4}{R_3 + R_4} u_{12} - \frac{R_2}{R_1} u_{11} \tag{2.27}$$

The usual choice is

$$R_1/R_3 = R_2/R_4 \tag{2.28}$$

Therefore the amplifier output voltage is proportional to the voltage difference on the input

$$u_2 = \frac{R_2}{R_1} (u_{12} - u_{11}) \tag{2.29}$$

This calculation was based on the ideal op-amp. In reality, the connection becomes more complex, since it is required to compensate offsets, drifts, sensitivity to common-mode voltage, etc. Therefore it is usually better to use directly a ready-made circuit, typically called an "instrumentation amplifier," which takes care of those problems.

2.5 Instrumentation amplifier

The instrumentation amplifier is a connection of at least three op-amps. The first two are used as buffers with high-input resistance. The input resis-

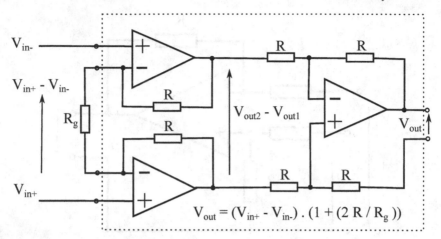

FIGURE 2.16: Instrumentation amplifier

tance is the same for both channels (This was not the case for the difference amplifier). The third op-amp is connected as a difference amplifier. The gain of the instrumentation amplifier is usually set with only one resistor, typically denoted as R_g. The schematic of the instrumentation amplifier is shown in figure 2.16. The resistors R are internal.

Different types of instrumentation amplifiers differ typically in gain and bandwidth.

3

Temperature—contact

Temperature is one of the most important properties. It influences virtually all other properties. Temperature is an integral property. It is a measure of average kinetic energy of the molecules.

There are different temperature scales. Absolute temperature is expressed in kelvin (K). By definition, the temperature at absolute zero is 0 K. The second point defining the thermodynamic scale is the triple point of water (0.01°C). In thermodynamics, the triple point of a substance is the temperature and pressure at which three phases (for example, gas, liquid and solid) of that substance coexist in thermodynamic equilibrium.

Other used temperature scales are Celsius, Fahrenheit and Rankine. The Celsius scale uses the freezing point (0°C) and boiling point (100°C) of water as definition points.

The formulas to recalculate between different selected scales are shown in table 3.1.

Defining points for temperature are nowadays defined by the International Temperature Scale (ITS-90) [7].

3.1 Resistive temperature detectors (RTD)

Resistive temperature detectors are based on changes of electrical resistance on temperature. They are among the most common industrial temperature sensors due to their accuracy, robustness and reliability.

The dependence of electrical resistance on temperature for some materials in a small temperature range can be considered linear. As small range ±100°C

TABLE 3.1: Conversion between selected temperature units

-	K	°C	F
K	-	$[K] = [°C] + 273.15$	$[K] = ([°F] + 459.67) \cdot (5/9)$
°C	$[°C] = [K] - 273.15$	-	$[°C] = ([°F] - 32)/1.8$
F	$[°F] = (9/5)x[K] - 459.67$	$[°F] = 1.8[°C] + 32$	-

is usually considered. The electrical resistance is

$$R_T = R_0[1 + \alpha T] \tag{3.1}$$

where R_0 is the electrical resistance for 0°C.

The temperature coefficient of resistance α is

$$\alpha = \frac{R_{100} - R_0}{100 \cdot R_0} \tag{3.2}$$

where R_0 is the electrical resistance for 0°C and R_{100} is the electrical resistance for 100°C.

The temperature coefficient of resistance α is the sensitivity, how much the electrical resistance is changing with changes of temperature. It is usually defined from the difference of resistances for 100°C and 0°C, as shown in equation 3.2. Temperature coefficients of resistance for metals common in industrial temperature sensors are shown in table 3.2.

The dependence 3.1 is linear, but with a good accuracy it is valid only in a small temperature range. For larger temperature ranges or for higher accuracy it is necessary to use a non-linear equation with more coefficients.

For example for a platinum RTD, the following equation is used

$$R_T = R_0[1 + AT + BT^2 + C(T - 100)T^3] \tag{3.3}$$

As can be seen from this equation, it is non-linear. However for a small range of temperature T, the contribution of the higher powers of T is negligible, and the equation approaches linearity.

For metallic RTDs, three metals are used almost excursively. They are platinum, nickel and copper. The properties of those elements significant for RTDs are summarized in table 3.2.

The most important property is the temperature coefficient of electrical resistance α. It tells how much is the electrical resistance changing with changes of temperature. It is also called sensitivity. From the table it is obvious, that the nickel sensor has roughly double sensitivity compared to platinum. It will however have other disadvantages, such as non-linearity as we will see later. Therefore it will be usable only in a small temperature range.

The second very important property is resistivity, shown in the last column of table 3.2. It expresses the electrical resistance of a wire with a given length and cross section. It can be seen that platinum and nickel have approximately

TABLE 3.2: Temperature coefficient of resistance α and resistivity ρ for selected metals used in RTDs

Element	α [10^{-3}/K]	ρ [Ω.m]
Pt	3,85	$9,81.10^{-8}$
Ni	6,17	$12,13.10^{-8}$
Cu	3,9	$1,54.10^{-8}$

the same resistivity. Therefore, the size of platinum and nickel sensors with the same resistance will be approximately equal. Copper, on the other hand, has about 6× smaller resistivity. Therefore the wire made from copper with the same diameter and cross section as a platinum wire will have significantly lower resistance. This will have an effect on the nominal resistance of RTDs made from different materials.

The most common nominal resistance (for 0°C) for platinum RTDs is 100 Ω; for copper it is 10 Ω.

3.1.1 Platinum RTDs

Platinum is a very stable metal. Platinum RTDs are among the most accurate and reliable temperature sensors because of the unique properties of platinum.

Platinum has a relatively high resistivity (compared to, e.g., copper), therefore the current passing through the sensor can be small. This will in turn prevent self heating of the sensors.

The biggest advantage of platinum sensors is, however, their accuracy and long-time stability. This allows calibrating the sensors and then relying on the calibration for a long period of time. Platinum RTDs are also easily interchangeable without re-calibration (as opposed to other sensor types).

Platinum RTDs are denoted with symbols Ptxxx, where xxx denotes the resistance for temperature 0°C.

The most common sensor, Pt100 has, therefore, resistance of precisely 100 Ω for 0°C.

Other very common type is Pt1000. It has precisely 1000 Ω for 0°C.

The resistance vs. temperature dependence for a platinum RTD in a large temperature range is

$$R_T = R_0[1 + AT + BT^2 + C(T - 100)T^3] \qquad (3.4)$$

coefficients A, B and C are defined by standard IEC 751 [8].

$$\begin{aligned} A &= 3,90832 \cdot 10^{-3} \quad K^{-1} \\ B &= -5,775 \cdot 10^{-7} \quad K^{-2} \\ C &= -4,183 \cdot 10^{-12} \quad K^{-4} \end{aligned} \qquad (3.5)$$

The sensitivity of a Pt100 is 0.385 Ω/°C [8].

Highest-purity platinum is used for RTDs (99,999%). Based on the purity and accuracy requirements (and of course price), there are two available tolerance classes. Class A and class B. Class A is more accurate (and more expensive), and class B is less accurate and less expensive. The tolerance classes are shown in table 3.3.

The platinum RTD is the most accurate temperature sensor in a wide temperature range. In fact, the International Temperature Scale uses the platinum RTD to define some fixed reference points. The usual temperature range for a Pt100 is −200°C to +600°C.

TABLE 3.3: Pt100 tolerance classes

Class	Tolerance
A	0.15+0.002|T|
B	0.30+0.005|T|

TABLE 3.4: Pt100—accuracy and tolerance classes

Temperature (−)	Class A (±°C)	Class A (±Ω)	Class B (±°C)	Class B (±Ω)
−200	0,55	0,24	1,3	0,56
−100	0,35	0,14	0,8	0,32
0	0,15	0,06	0,3	0,12
100	0,35	0,13	0,8	0,3
200	0,55	0,20	1,3	0,48
300	0,75	0,27	1,8	0,64
400	0,95	0,33	2,3	0,79
500	1,15	0,38	2,8	0,93
600	1,35	0,43	3,3	1,06

For example for tolerance class A and for temperature 0°C it is possible to achieve accuracy of ±0, 15°C, for temperature 600°C accuracy ±1, 35°C is achievable.

Other temperatures and tolerance-class B is shown in table 3.4.

Due to the different standards in purity as defined by DIN and IEC or ANSI, there are different sensitivities available on the market. DIN and IEC = sensitivity 0,3850 Ω/°C, ANSI = sensitivity 0,3923 Ω/°C.

This limits interchangeability of the platinum RTD. The sensor resistance for 0°C is exactly the same, but the difference in sensitivity yields different steady-state characteristics. This leads to a different resistance value for higher (and lower) temperature. The significance of this difference is shown in figure 3.1 and in table 3.5. It is clearly visible that for example, 600°C the difference is -3,88 Ω. This is approximately 10°C, about 9× larger than the accuracy. The choice of a different standard from the one used previously in an application will have an extreme impact on accuracy. This difference is not at all saying that one standard is better than the other one. Not at all. It is just saying that the user should be aware of the difference.

3.1.1.1 Glass platinum RTD

RTD's are common in three different types—glass, ceramics and thin film. The most common out of those three is glass. Its main components are shown in figure 3.2.

The sensor is composed of a thin platinum wire with a diameter of approximately 0.05 mm. The wire is encapsulated in a glass casing. The wire winding is bifilar, in order to minimize undesirable inductance. The wire is lead into

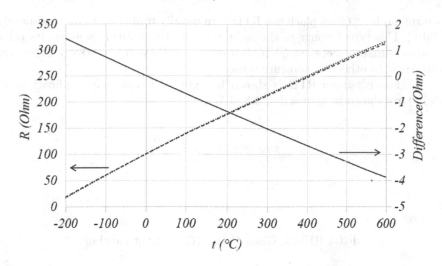

--- *R IEC751 (Ohm)* ······ *R ASTM E1137 (Ohm)* —— *Difference (Ohm)*

FIGURE 3.1: Pt100—DIN, IEC 751:2008 [8] vs. ANSI ASTM E1137 [9]

TABLE 3.5: Pt100—DIN, IEC vs. ANSI

Temperature (°C)	R IEC751 [8](Ω)	R ASTM E1137 [9](Ω)	Δ (Ω)
−200	18,52	17,08	1,44
−100	60,26	38,65	1,07
0	100	100	0
100	138,51	139,2	−0,69
200	175,86	177,23	−1,7
300	212,05	214,08	−2,03
400	247,09	249,76	−2,67
500	280,98	284,26	−3,28
600	313,71	317,59	−3,88

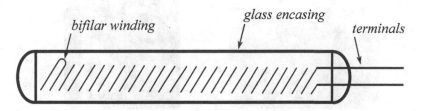

FIGURE 3.2: Glass Pt100

a terminal box. Glass platinum RTDs are usually usable up to approximately 600°C. This type belongs to the most accurate temperature sensors. Its price is approximately between 50 to 100 euros, so they are relatively expensive compared to other temperature sensors.

A glass platinum RTD is shown in figure 3.3. Example applications of Pt100 are shown in figures 3.4, 3.5, 3.6, 3.7.

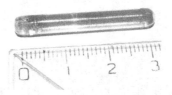

FIGURE 3.3: Glass Pt100 RTD—photo author

FIGURE 3.4:
Pt100—measurement of water temperature from a boiler—photo author

FIGURE 3.5:
Pt100—measurement of temperature of exhaust gases—photo author

FIGURE 3.6:
Pt100—measurement of temperature of exhaust gases—photo author

FIGURE 3.7:
Pt100—measurement of temperature in a tank during beer production—photo author

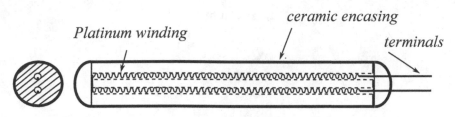

FIGURE 3.8: Ceramic Pt100

RTDs are quite fragile, especially the glass type. All RTDs must therefore be protected from mechanical and chemical damage with a thermowell. Thermowells will be discussed in a separate section.

3.1.1.2 Ceramic platinum RTD

FIGURE 3.9: Ceramic Pt100—photo author

The second type of platinum RTD is made from ceramics. The construction is very similar to the glass type, but the casing is ceramics. The platinum wire is not melted into the material, but it is placed in a tube. The arrangement of the ceramic platinum RTD is shown in figure 3.8. Ceramic platinum RTDs are shown in figure 3.9.

The ceramic platinum RTD has smaller hysteresis, but smaller resistance to vibrations. They are often used as a temperature standard of higher order in metallic thermowells filled with air or a helium oxygen mixture and hermetically sealed. A direct contact with atmosphere causes higher measurement uncertainty cased by the influence of hydrogen and carbon dioxide.

The wire is allowed to move freely inside the casing. This limits the additional strain effect, caused by the different thermal expansion of glass (approx. $6 \cdot 10^{-6}(m/(mK))$) and platinum (approx. $9 \cdot 10^{-6}(m/(mK))$).

The wire diameter is about 0.05 mm. The sensor has to be placed in a thermowell. Thanks to the ceramic casing, the sensor is suitable for higher temperatures, up to approximately 850°C.

1,6 x 3,2 mm

FIGURE 3.10: Thin film Pt1000—photo author

3.1.1.3   Thin-film RTDs

The third construction type is thin film. In the manufacturing process a thin film of platinum paste is deposited on a ceramic substrate (Al_2O_3). Since the used amount of platinum is very small, the sensor is a lot cheaper that the previous types.

The platinum thickness is a few µm. The layer is then baked and laser trimmed to the desired resistance. Thin-film sensors are very small, they have a fast response (small time constant). Due to their small size, they also have less influence on the measured surface. The typical use for thin film sensors is the measurement of surface temperature. A thin film Pt100 is shown in figure 3.10.

3.1.2   Nickel RTDs

The biggest advantage of platinum RTDs shown in the previous chapter is their accuracy. They are however expensive. For low cost applications, where accuracy is not an issue, Nickel RTD's can be used. The typical application where Nickel RTDs are used is for example household temperature controllers. In those applications, accuracy of about ±1°C is sufficient.

Nickel is cheap when compared to platinum. Those sensors are therefore significantly less expensive that platinum RTDs. They are however also much less accurate and less stable in the long-term and are non-linear.

Compared to platinum, nickel has about double the sensitivity. A nickel sensor is therefore more sensitive, and the electrical resistance changes more with temperature.

The relation between electrical resistance and temperature is as follows

$$R_T = R_0[1 + AT + BT^2 + CT^4 + DT^6] \qquad (3.6)$$

The equation and coefficients are based on the standard DIN 43760. The values of the coefficients are shown in table 3.6. It can be seen that the equation is non-linear.

Nickel RTDs are commonly used for temperature up to 200°C. The naming conventions are the same as for platinum. The most common type is Ni120. It has resistance of 120 Ω for 0°C.

Standard DIN 43760 defines two tolerance classes, A and B, as shown in table 3.7. Figure 3.11 compares steady-state characteristics of Pt100, Ni120 and Cu10.

TABLE 3.6: Coefficients for nickel resistor according to DIN 43760 [10]

Coefficient	value
A	$5.481 \cdot 10^{-3}$
B	$6.650 \cdot 10^{-6}$
C	$2.805 \cdot 10^{-11}$
D	$2.000 \cdot 10^{-17}$

TABLE 3.7: Tolerance classes for Nickel RTDs

Class	+/− accuracy (°C)	
	t < 0 °C	t > 0 °C
A	$0.2+0.014|T|$	$0.2+0.0035|T|$
B	$0.4+0.028|T|$	$0.4+0.007|T|$

From figure 3.11 it can be seen that Pt100 is linear in the shown range; Ni120 is not. Also the higher sensitivity for Ni120 is visible. For the Cu10 RTD, the smaller initial resistance (10Ω for 0°C) can be seen. The sensor is linear, it has approximately the same sensitivity (slope) as Pt100.

Nickel RTDs have a similar construction as platinum sensors. The most common type is thin-film. The price of a nickel RTD is approximately 20 euro/piece (about 5× lower that Pt100).

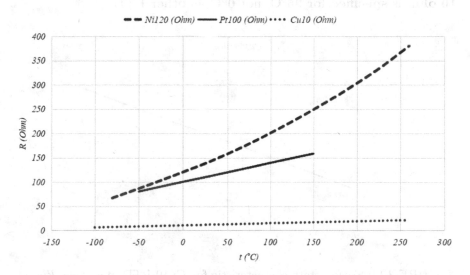

FIGURE 3.11: Steady-state responses for Pt100, Ni120 and Cu10

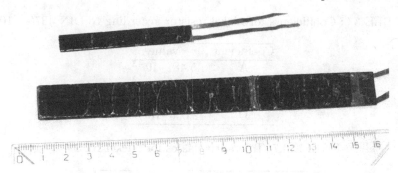

FIGURE 3.12: Cu10 RTD—photo author

3.1.3   Copper RTDs

Copper RTDs can be used for short-time experiments. Copper reacts strongly with the atmosphere, the sensor changes its properties in time. There are however applications where this is not an issue. One example is measurement of winding temperature in electric machines. The winding temperature is measured, and the sensor is removed. No long-time stability is required. Since copper is significantly cheaper that platinum or nickel, the sensor is low cost. An example of a Cu10 copper RTD for measurement of temperature in electric machines is shown in figure 3.12. In the used temperature range, the steady-state characteristic is linear as shown in figure 3.13. **Note that $R = 10$ ohm is specified for 25°C, not 0°C as other RTDs.**

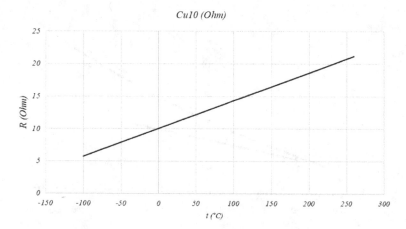

FIGURE 3.13: Steady-state characteristic for Cu10 RTD, note that $R = 10$ ohm is specified for 25°C, not 0°C as other RTDs

3.2 Thermistors

Thermistors are semiconductor components. They are produced by high-temperature pressing of metallic oxides ($Fe_2 O_3$, TiO_2, CuO_2, NiO) into a form of a bead. They exploit the dependence of electrical resistance on temperature. Their behavior is in some aspects similar to RTDs. However the dependence of resistance on temperature for semiconductors is non-linear. Depending on the semiconductor material, the resistance may decrease with temperature (NTC—Negative Temperature Coefficient) or increase (PTC—Positive Temperature Coefficient). The resistance-temperature dependence for both types is shown in figure 3.14.

In both cases the dependence is non-linear. Moreover, for PTC the resistance typically first drops, and then there is a sharp increase. The sharp increase occurs above the Curie temperature; the resistance can increase more than three orders of magnitude. NTCs are typically used as temperature sensors. Their resistance-temperature dependence is non-linear but monotone. PTCs are typically used as resettable temperature fuses, although temperature sensors can be built with PTCs as well.

In most cases the resistance-temperature dependence is approximated with the Steinhart-Hart equation 3.7

$$\frac{1}{T} = A + B \ln R + C(\ln R)^3 \tag{3.7}$$

As can be seen from the equation, the dependence is logarithmic. The inversed dependence, resistance on temperature is exponential. Coefficients A, B and C in the equation are dependent on the used semiconductor material.

$$R = e^{(Y - \frac{X}{2})^{1/3} - (Y + \frac{X}{2})^{1/3}} \tag{3.8}$$

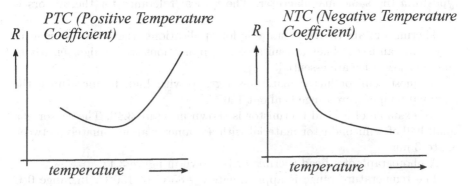

FIGURE 3.14: Resistance-temperature dependence for NTC and PTC

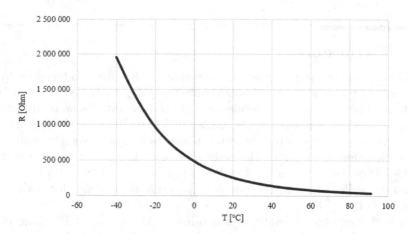

FIGURE 3.15: Example of steady-state characteristics for thermistor

where

$$X = \frac{1}{C}(A - \frac{1}{T})$$ (3.9)

$$Y = \sqrt{(\frac{B}{3C})^3 + \frac{X^2}{4}}$$ (3.10)

An example of the resistance-temperature dependence for a thermistor with $20k\Omega$ for 25°C is shown in figure 3.15.

Thermistors have a large sensitivity. It is approximately $10\times$ larger than for metal RTDs. They are, however, non-linear. Therefore, a more complex electronic circuit, typically a microcontroller, is used to handle this. The microcontroller is used to linearize the response.

Another disadvantage of thermistors is their large tolerance. In practice, it means that sensors needs to be calibrated when replaced even with the same type from the same manufacturer. The typical tolerance for thermistors is about 5% to 10%.

Thermistors are therefore suitable for applications where the lower accuracy is not an issue. They are suitable for applications where high sensitivity and very low price are essential.

As most semiconductor materials degrade with high temperatures, the maximal temperature range is about 150°C.

An example of a bead thermistor is shown in figure 3.16. The sensor is a small ball of semiconductor material with a diameter approximately between 0,2 to 3 mm.

A photograph of a bead thermistor is shown in figure 3.17.

The temperature range is approximately −85°C to 150°C. In range 0°C to 70°C the typical accuracy is ±0,2 °C. In larger temperature ranges, for example −55°C to 150°C, the typical accuracy is ±0,4°C.

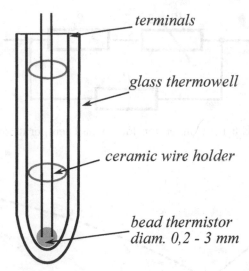

terminals

glass thermowell

ceramic wire holder

*bead thermistor
diam. 0,2 - 3 mm*

FIGURE 3.16: Bead thermistor in a thermowell

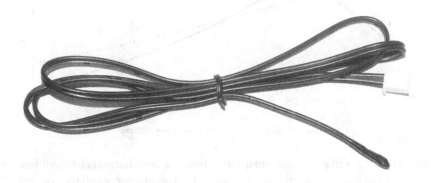

FIGURE 3.17: Bead thermistor—photo author

The biggest advantage is the low price compared to platinum RTDs. The price is about 1 order of magnitude lower. On the other hand, the accuracy is also about 1 order of magnitude smaller.

Thermistors are very fast; they have a small time constant. It is typically about 1 second for the sensor without a thermowell. With a thermowell, the time constant will increase. Thermistors are therefore suitable to measure fast temperature transients.

Thermistor linearity Since the thermistor is a non-linear sensor, it can either be used only in a very small temperature range or the steady-state characteristic needs to be adjusted. In a very small range, the characteristic can be made linear in an "electric" way, with a serial and parallel resistor. An example for a thermistor for 25°C is shown in figure 3.18.

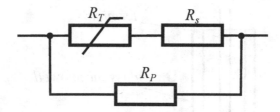

FIGURE 3.18: Connection for a small temperature range

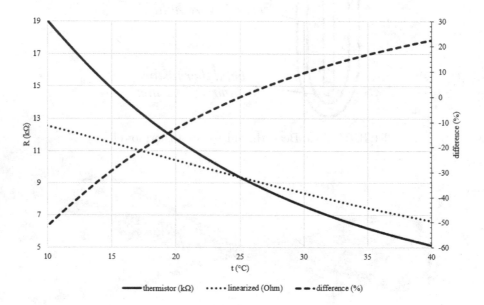

FIGURE 3.19: Original and linearized steady-state characteristic, difference
of both in %, simulation prepared by the author, available on
www.circuitlab.com/circuit/7r27s8/termistor/

There are two resistors connected to the thermistor R_t. In this example,
the thermistor has resistance of 10 $k\Omega$ for 25°C. The resistor R_s is connected
in series; the resistor R_p is connected in parallel. A good choice in this case is
$R_s = R_p = 10\ k\Omega$.

The result is shown in figure 3.19. Around the operating point, 25°C, the
difference is small. As the temperature is changing, the difference gets bigger.
This connection is good for about ±5°C. For higher range, the differences
would be too large.

Another possibility for how to handle the non-linearity is to use a look-
up table stored in a microcontroller. The microcontroller is doing piece-wise
interpolation between the stored points. The block diagram is shown in figure 3.20. The microcontroller output is usually a standard voltage or current

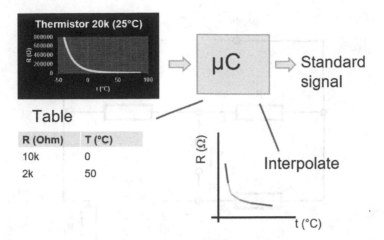

FIGURE 3.20: Piece-wise interpolation with micro-controller

signal. The advantage of this approach is the capability to work in a whole temperature range of the thermistor. On the other hand, the electronic circuit is significantly more complex. Due to lower prices of micro-controllers, this is currently the preferred approach.

3.3 Circuits for RTDs

The change of electric resistance for RTDs is usually very small. For example, Pt100 has a sensitivity of $0,39\ \Omega/°C$. Therefore the change of resistance is not measured directly. Special circuits are used. For RTDs, bridge circuits are used. As we are looking for electric resistance, a DC variable, a DC bridge will be used.

3.3.1 Two-wire connection

This connection is the simplest possible. A direct two-wire connection is made between the bridge and the sensor. The connection is shown in figure 3.21. However, the connection does not compensate for the eventual change of resistance of the connecting wires. This can occur with a change of ambient temperature. This connection is therefore suitable only for applications where the ambient temperature is not changing or the connection wires are very short.

The circuit simulation where this can be tested is prepared at https://www.circuitlab.com/circuit/p99q68/temp-sensor-2-wire-connection/.

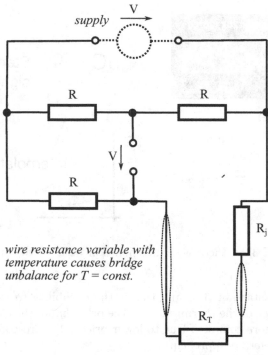

sensor temperature T

FIGURE 3.21: Two-wire connection for RTD

The resistor R_j is used to set a specific value of resistance of the connection. Its typical size is 16 Ω or 20 Ω.

The influence of the change of resistance on the bridge output voltage is shown in the following example. The TP Ethernet cat 5e wire is used. It is used not because it is suitable for sensors, but because it is known to most readers. The wire resistance is approximately $R \approx 0,19$ Ω/m.

The temperature coefficient of resistance of copper is $\alpha = 4 \cdot 10^{-3} 1/K$. Let's have a wire length of 10 m, change of ambient temperature of 10°C. The wire will change its electrical resistance by 80 mΩ (to the sensor and back).

For a Pt100 RTD with sensitivity of 0,385 Ω/°C, this represents an additional error of approximately 0.2°C.

3.3.2   Three-wire connections

The three-wire connection is used in applications where temperature compensation is required. This will apply also for longer wires. The connection is shown in figure 3.22.

The simulation of the circuit is available at
https://www.circuitlab.com/circuit/8k84bb/temp-sensor-3-wire-connection/.

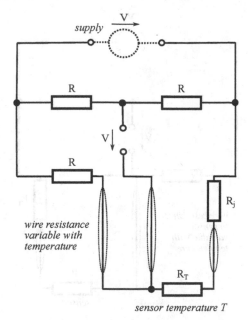

FIGURE 3.22: Three-wire connection for RTD

It can be seen that the change of ambient temperature has no effect on the bridge output voltage. The change is compensated.

The bridge is initially balanced. When there is a change of the ambient temperature, the resistance of the connecting wires is changing. The wires are, however, connected in both diagonals of the bridge. By looking at the condition of balance it can be seen that the change of ambient temperature is compensated. This assumes that all three connecting wires change temperature (and resistance) in the same way.

The resistor R_j is used to set a specific value of resistance of the connection. Its typical size is 16 Ω or 20 Ω.

3.3.3 Two-wire with compensating loop

This is a four-wire connection. It is shown in figure 3.23.

The bridge is connected with two wires to the sensor. There are two additional wires connected at the sensor's terminal box. The wires lead the same way to the sensor. Therefore they will also change their temperature in the same way. The wires are connected to the opposite diagonal of the bridge. The change of electrical resistance caused by ambient temperature change is compensated.

The resistor R_j is used to set a specific value of resistance of the connection. Its typical size is 16 or 20 Ω.

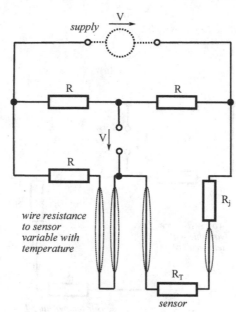

FIGURE 3.23: Two-wire connection with compensating loop

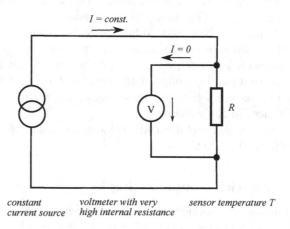

FIGURE 3.24: Four-wire connection

3.3.4   Four-wire connections

This connection, shown in figure 3.24 is more suitable for thermistors. They have a larger change of resistance, and therefore a larger signal is produced.

The connection requires a current source. This source provides a constant current, regardless of the resistance. The voltage is measured with a voltmeter with a high-input resistance. Since the voltmeter has a high-input resistance, the current flowing through the voltmeter is small, ideally zero. Therefore, there will be no voltage drop on the wires between the sensor and the voltmeter. Therefore the change of wire resistance will not have an effect on the measured voltage.

3.3.5 Sensor self-heating

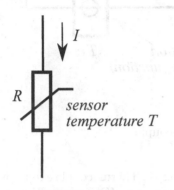

The sensor will self-heat with the current, as shown in figure 3.25. Therefore the current needs to be limited. The typical maximal current for a Pt100 is 1 mA; for thermistors it is about 50 µA. The shown currents lead to Joule losses at the sensor of approximately 100 µW.

The heat produced in the sensor is

$$P = R \cdot I^2 \qquad (3.11)$$

FIGURE 3.25: Sensor self-heating

3.4 Thermocouples

Thermocouples are based on the Seebeck effect. They are created as a conductive connection of two dissimilar metals. A schematic connection of a thermocouple is shown in figure 3.26; the thermocouple wire with a welded hot end is shown in figure 3.27.

Thermocouples are produced by welding. The thermocouple wire is welded at the hot junction (hot end). This end will be used to measure. The other end is called the cold (reference) junction. The thermoelectric voltage measured between the two open wires at the cold end is proportional to temperature difference. The voltage depends on the thermocouple wire materials.

A thermocouple measures temperature difference. This is different from all other sensors.

If the absolute temperature should be measured, the cold junction temperature has to be known. As for all the other sensors, a thermocouple is typically placed in a thermowell. The thermocouple wire then ends in a terminal box at the end of the thermowell. From here the wire can continue either as a thermocouple wire or as a compensation wire.

Thermocouple wire from the terminal box will use the same thermocouple wire as the main thermocouple. The advantage is that such a connection

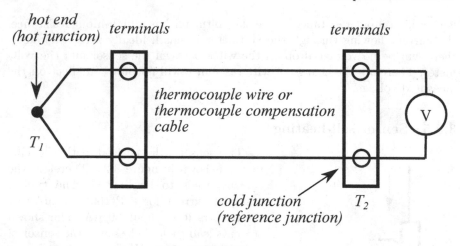

*hot end
(hot junction)* *terminals*   *terminals*

*thermocouple wire or
thermocouple compensation
cable*

T_1

V

*cold junction
(reference junction)* T_2

$$V \sim \Delta T = T_1 - T_2$$

FIGURE 3.26: Thermocouple

has exactly the same properties as the thermocouple. Thermocouple extension cable is identified by the letter **X** in the cable type, e.g., KX for type-K thermocouple. However for high-temperature thermocouples, made from expensive metals such as platinum, the wire would be too expensive for long connections. Therefore the compensation wire is used.

Compensation wire has similar electrical properties as the thermocouple wire, but only in a limited temperature range (typically to 100°C or 200°C). It is made from less expensive metals and is therefore less expensive. The compensating cable is identified by a letter **C** (e.g., KC for type-K thermocouple).

The thermocouple type is letter- and color-coded. The most common thermocouple types are type J and type K.

Thermocouple Type J is made from iron and constantan. Constantan is an alloy of copper and nickel.

Thermocouple type K is made from an alloy of nickel and chromium and nickel and aluminum.

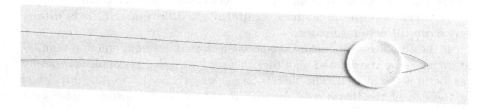

FIGURE 3.27: Thermocouple—photo author

TABLE 3.8: The most common thermocouple types

Type	Material	sensitivity in range 0 to 100°C
J	Fe + Constantan (CuNi)	52,69 µV /100°C
K	NiCr + NiAl	40,96 µV / 100°C

The properties of those two most common thermocouples are summarized in table 3.8. Both thermocouple types are used for example in handheld multimeters. The temperature range depends on the construction but is typically few hundred°C.

The color coding of selected thermocouple types is shown in table 3.9 and their properties are shown in table 3.10. Steady-state characteristics of selected thermocouples, based on data from [11], colors according to IEC are shown in figure 3.28.

3.4.1 Thermocouple installation

Figure 3.29 shows three different types of thermocouple installation.

The bare thermocouple does not have any mechanical or chemical protection. It is unprotected. On the other hand, it has the fastest response.

The isolated thermocouple is placed in a thermowell. It is electrically insulated from the thermowell. The thermowell provides a very good protection. The thermowell increases the response time (time constant) of the sensors.

The grounded thermocouple is electrically connected to the thermowell. It is, for example, welded or soldered. This provides good heat transfer. The sensor is faster than the isolated version. On the other hand, the grounding may cause trouble with electromagnetic interference. A thermocouple probe is shown in figure 3.30.

TABLE 3.9: Materials and colors for selected thermocouples—based on [12], [13], [14], [15], [16], [17], [18], [19]

Type	Materials	Cable color (IEC)	Cable color (ANSI)	Wire colors (IEC)
J	Fe, CuNi	black	black	white (+), red (−)
K	NiCr, NiAl	green	yellow	yellow (+), red (−)
T	Cu, CuNi	brown	blue	brown (+), white (−)
E	NiCr, CuNi	magenta	magenta	magenta (+), white (−)
N	NiCrSi, NiSiMg	pink	orange	pink (+), white (−)
R	Pt 13% Rh, Pt	orange	green	orange (+), white (−)
S	Pt 10% Rh	orange	green	orange (+), white (−)

TABLE 3.10: Properties of selected thermocouples—based on [20], [12], [13], [14], [15], [16], [17], [18], [19]

Type	Usual range (°C)	Sensitivity 0 to 100°C (μV/°C)	Accuracy (whichever is larger)	Use
B	0 to 1700	33	0,5°C	Oxidizing or inert atmosphere, do not use in metallic thermowells, used in glass industry
J	0 to 750	52,69	2,2°C or 0,75%	Reducing atmosphere, vacuum, inert, limited use in oxidizing atmosphere for higher temperatures, not recommended for low temperatures
K	−200 to 1250	40,96	2,2°C or 0,75% above 0°C, 2,2°C or 2% below 0°C	Clean oxidizing and inert atmosphere, limited use in vacuum or reducing atmosphere
T	−200 to 350	42,79	1,0°C or 0,75% above 0°C, 1,0°C or 1,5% below 0°C	Slightly oxidizing, reducing, inert atmosphere, vacuum, suitable in presence of humidity, for low temperatures and cryogenic applications
E	−200 to 900	63,19	1,7°C or 0,5% above 0°C, 1,7°C or 1,0% below 0°C	Oxidizing or inert atmosphere, limited use in vacuum or reduction atmosphere
N	−270 to 1300	27,74	2,2°C or 0,75% above 0°C, 2,2°C or 2,0% below 0°C	an alternative for type K, more stable for higher temperatures
R	0 to 1450	6,47	1,5°C or 0,25%	Oxidizing or inert atmosphere, do not use in metallic thermowells
S	0 to 1450	6,46	1,5°C or 0,25%	Oxidizing or inert atmosphere, do not use in metallic thermowells

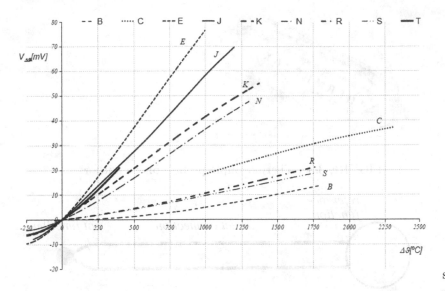

FIGURE 3.28: Steady-state characteristics of selected thermocouples, based on data from [11], colors according to IEC

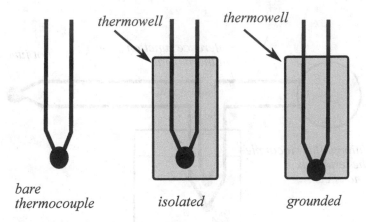

FIGURE 3.29: Thermocouple installation—bare thermocouple, isolated and grounded

As shown in the previous paragraphs, the thermocouple measures **temperature difference**. If the cold junction temperature is changing, the voltage changes as well. Therefore the cold junction temperature changes needs to be compensated. It can be maintained constant with a thermostat as shown in figure 3.31.

The thermostat can be filled with a water–ice mixture, maintaining a constant temperature of 0°C. It can also be filled with oil and heated, for example to 50°C. The temperature is then maintained with a temperature controller.

FIGURE 3.30: Thermocouple probe—photo author

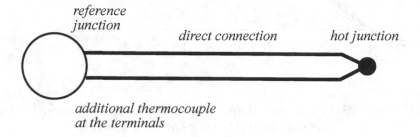

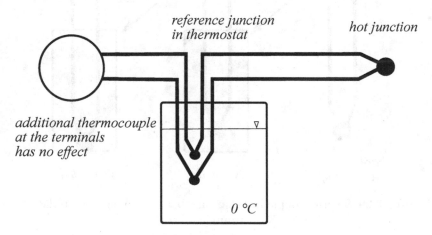

FIGURE 3.31: Thermostat—maintains constant cold-end temperature

3.4.2 Cold-junction box

The cold-junction temperature changes may also be compensated with a cold-junction box. It is a bridge circuit, balanced for certain temperature. Typically 20°C. The connection is shown in figure 3.32, a real cold-junction box is shown in figure 3.33.

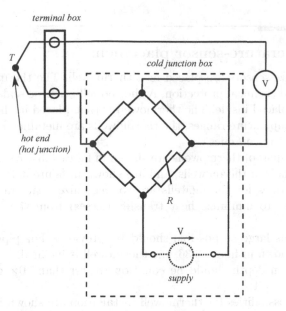

FIGURE 3.32: Principle of cold-junction box

FIGURE 3.33: Cold-junction box—photo author

The bridge is composed from three resistors independent on temperature (made from constantan, $\alpha = -0,000074\ 1/°C$). The fourth resistor is made from copper ($\alpha = 0,004041\ 1/°C$) and is therefore variable with temperature. The bridge is balanced for 20°C. If the ambient temperature is changing the bridge gets unbalanced. The bridge voltage is added to or subtracted from the thermocouple voltage.

One cold-junction box is required for each thermocouple. This connection is therefore suitable only when just a few thermocouples are use. For a larger number of thermocouples, the thermostat is a more suitable solution. All thermocouples can use the same thermostat.

3.5   Temperature-sensor placement

Temperature sensors are placed in a thermowell. The thermowell assures mechanical and chemical protection. A thermowell is a pipe sealed at one end. The sensor is placed inside. The thermowell is then placed in the application. For smaller temperature ranges, the thermowells are metallic; for higher temperatures they are ceramics.

A few examples of thermowells are shown in figure 3.34. Examples include industrial Pt100's in thermowells, and are shown in figure 3.35

The main rules for thermowells are to maximize heat transfer into the thermowell and to minimize heat transfer (losses) from the thermowell to ambient.

A surface as large as possible should be provided. For pipes with small diameter, as shown in figure 3.36, the thermowell is inclined.

The insertion depth should be equal or greater than 10× external thermowell diameter [21].

Different possibilities for thermowell installation are shown in figure 3.37.

From the heat transfer point of view, the best position is in figure 3.37 A. The thermowell is in the elbow, against flow direction. The recommended distance from heat exchangers, mixers, etc., is at least 25× pipe diameter [21]. This assures a good mixing of the liquid in the elbow, and a good heat transfer.

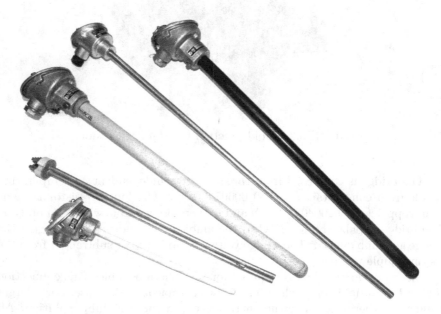

FIGURE 3.34: Thermowells—photo author

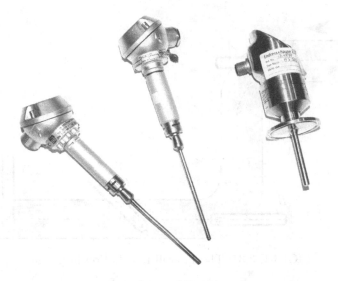

FIGURE 3.35: Industrial Pt100—photo author

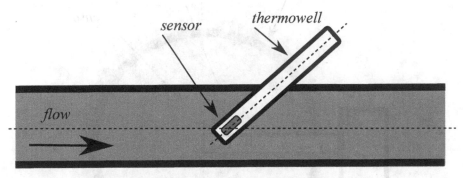

FIGURE 3.36: Thermowell in a small pipe, inclined against flow direction

Position shown in 3.37 B, in the elbow, again flow direction is not optimal. The heat transfer is smaller in this area because of low or no turbulence. The heat transfer is not as good as in position A.

The inclined thermowell is shown in figure 3.37 C. The inclination increases the surface area and the heat transfer.

The installation is also possible in a T-piece as shown in figure 3.37 D. The insertion depth is about 0,5 Φ D. The insertion depth can be larger if an increase in the surface area is required. Longer thermowell, on the other hand, increases the risk of Von Karmán vortices; this can cause vibrations and fatigue of the thermowell. For high-flow velocities, it is therefore recommended to use a shorter thermowell.

In large-diameter pipes, the thermowell is installed without inclination, perpendicular to the pipe axis as shown in figure 3.38.

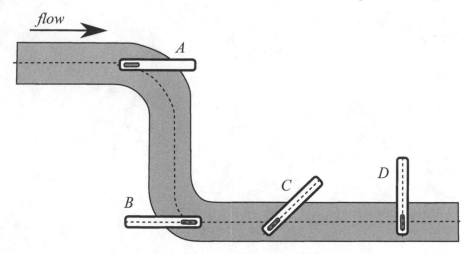

FIGURE 3.37: Thermowell installation in pipes

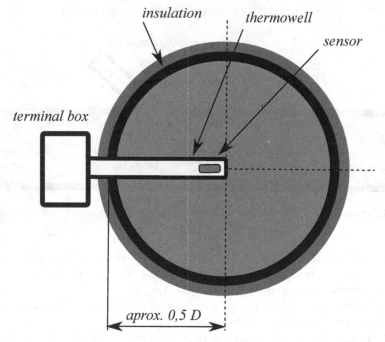

FIGURE 3.38: Thermowell in pipe with a large diameter

The recommended insertion depth is 0.5 D. For fast flow rates, it should be shorter.

An example is shown in figure 3.38. It includes insulation to prevent heat losses.

FIGURE 3.39: Thermowells   FIGURE 3.40: Thermostat with
                              mixing—photo author

For smaller pipe diameters, the installation into an elbow or T-piece is
common.

Example of thermowells in a tank is shown figure 3.39. For sensor installation in tanks, the fluid mixing is very important. Mixing provides a uniform
temperature field. If mixing is not allowed, the temperature in the tank will
be unevenly distributed as shown in figure 3.41. Mixing, on the other hand,
may cause trouble with proper liquid-level sensing.

To assure an uniform temperature profile, thermostats with mixing are
available, an example is shown in figure 3.40.

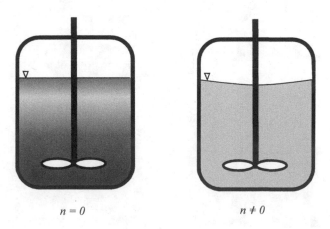

$n = 0$ $\qquad\qquad\qquad\qquad$ $n \neq 0$

FIGURE 3.41: Left—temperature profile in a tank without mixing,
right—temperature profile with mixing, unsteady liquid level

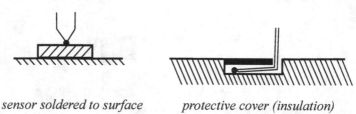

sensor soldered to surface   *protective cover (insulation)*

FIGURE 3.42: Temperature sensor for solid objects

For solid objects, the heat conductivity is important. For objects with good heat conductivity, the temperature field will not be distorted as much with the sensor installation. Therefore, the sensor can be mounted directly to the object's surface, or soldered or welded on top of it as shown in figure 3.42 left. In the case of bad heat conductors, heat losses through the sensor need to be prevented. The sensor is therefore covered with an insulation layer as shown in figure 3.42 right.

4

Temperature—non-contact

Non-contact temperature sensors are used to measure objects that are too far from the sensor or their temperature is outside the measurable temperature range of contact sensors. One example is the temperature measurement in furnaces. It is important to realize that non-contact temperature sensors measure intensity of radiation, not the temperature itself. Every object with a temperature above absolute zero radiates electromagnetic radiation according to Planck's law

$$I(v,T) = \frac{2hv^3}{c^2} \frac{1}{e^{\frac{hv}{kT}} - 1} \tag{4.1}$$

where h is Planck's constant $(6{,}62606896(33)\cdot10^{-34}$ Js$)$, c is speed of light in vacuum $(299\ 792\ 458$ m/s$)$ and k is Boltzmann constant $(1{,}380\ 648\ 52\ \cdot10^{-23}$ J/K$)$.

The intensity of the emitted radiation is a function of absolute temperature.

There are, however, other components when performing non-contact temperature measurement. They are shown in figure 4.1.

One of the important components is ambient radiation. It can be directed directly onto the sensor and on the measured object as well. Another

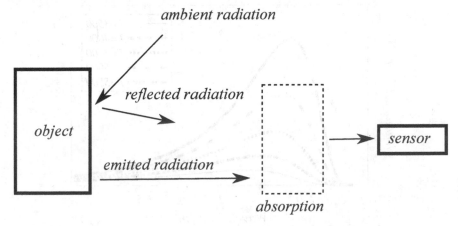

FIGURE 4.1: Components of radiation for non-contact temperature measurement

component is the reflected radiation. It is reflected from the object onto the sensor but is not related to the object's temperature.

For transparent objects, there can be also transmitted radiation from sources behind the measured object.

When the distance between the measured object and sensor is large, the absorption in the atmosphere will also play an important role. It will also play a significant role in environments with high relative humidity. High water-vapor content absorbs infrared (IR) radiation significantly.

All the aforementioned components cannot be distinguished at the sensor. Since the object's temperature is proportional only to its emitted radiation, various ways to limit the other components will be described.

As it is obvious from figure 4.2, the peak of maximal radiation intensity is shifting to lower wavelengths with increasing object temperature.

The non-contact temperature measurement can therefore be based either on the measurement of total emitted energy or on the intensity measurement for some selected wavelength.

The former is used in total radiation pyrometers, the latter in monochromatic pyrometers. Total radiation pyrometers measure basically the area under the curve shown in figure 4.2. Monochromatic pyrometers measure the intensity for a selected wavelength, in figure 4.2 shown with an example at 650 nm (red color).

For temperatures below 1000°C an important part of the radiated energy is in the IR part of the spectrum, as shown in table 4.1.

When working with non-contact thermometers the key term is emissivity ϵ. It is defined as the relative power of a surface to emit heat by radiation; the ratio of the radiant energy emitted by a surface to that emitted by a black

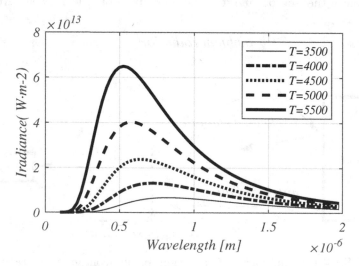

FIGURE 4.2: Black body radiation

TABLE 4.1: Approximate wavelength where 90% of all energy is radiated - based on [22]

Temperature (°C)	approx. wavelength (μm)
1000	1.5–8
500	3–16
200	4.5–30
100	5–40
0	8–45
−100	12–75

body at the same temperature. The user needs to be aware that a non-contact thermometer does not measure temperature but radiated energy. In order to show temperature, the emissivity of the measured object has to be known and correctly set on the thermometer. The purpose of this experiment is to estimate emissivity of different surfaces and to estimate the error caused by the wrong setting of emissivity (or fixed—some IR thermometers don't allow to change emissivity).

For all non-contact temperature measurements the most important parameter is emissivity. It is defined with the help of an ideal black body object. The black body is defined as an object that absorbs all incoming radiation of all wavelengths. At the same time it emits the maximal possible energy out of all possible objects with the same temperature.

Few examples of emissivity for selected materials are summarized in table 4.2.

As it is evident from the table, emissivity varies in a broad range. It is dependent on material, surface, color and temperature. Therefore it will be important to set the correct emissivity in accordance with the wavelength used for the measurement.

The radiation of a black body is described by Planck's law. It relates the spectral radiance of a black body (the quantity of radiation) with its temperature and wavelength of the radiation. A black body emits total radiant power W_B into a surrounding hemisphere given by [23]

$$W_B = \sigma \cdot T^4 \tag{4.2}$$

Where σ is Stefan-Boltzman constant and T is temperature in Kelvin.

TABLE 4.2: Emissivity of selected materials

material	emissivity
Al - non-oxidized	0,12–0,18
Fe shiny	0,28–0,42
Cu	0,1–0,35
Graphite, coal	0,65–0,97
Ideal black body	1

Any other body can be characterized by a dimensionless parameter–emissivity

$$\epsilon = \frac{W}{W_B} \tag{4.3}$$

It is the fraction of black body power emitted in the surrounding hemisphere. Emissivity depends on the surface of the body and on its temperature. By definition it is 1 for a black body. The black body is an idealized concept. Real objects do not absorb all incident energy; some part is reflected. They behave like gray bodies. Theirs emissivity is $\epsilon < 1$.

In order to correctly measure temperature with an IR thermometer, the emissivity of the measured object has to be known.

When the surface emissivity is not known, the following procedure can be used.

The real surface temperature can be measured with a contact thermometer, with the IR thermometer, and then emissivity can be calculated

The example will be shown for a constant emissivity set on the IR thermometer to $\epsilon = 0.95$. The temperature for this emissivity will be noted with symbol $T_{0.95}$.

The IR thermometer measures the radiated energy W_{IR}

$$W_{IR} = \epsilon_{0.95} \cdot \sigma \cdot T^4{}_{0.95} \tag{4.4}$$

Where $T_{0.95}$ is the temperature shown on the IR thermometer with fixed emissivity 0,95.

The object's radiated energy is a function of its temperature T_{obj}.

$$W_{obj} = \epsilon_{obj} \cdot \sigma \cdot T^4{}_{obj} \tag{4.5}$$

In order to determine object emissivity ϵ_{obj}, the object temperature T_{obj} is measured with a contact thermometer.

Then the object emissivity ϵ_{obj} can be calculated as

$$\epsilon_{obj} \cdot T^4{}_{obj} = \epsilon_{0.95} \cdot T^4{}_{0.95} \tag{4.6}$$

$$\epsilon_{obj} = \epsilon_{0.95} \frac{T^4{}_{0.95}}{T^4{}_{obj}} \tag{4.7}$$

The temperatures have to be substituted in Kelvin.

The emissivity changes with object's temperature and with the used wavelength as shown in [24] on the example of a SiC composite. Therefore for monochromatic pyrometers it will be very important to match the emissivity setting for the used wavelength.

The emissivity setting is therefore very important. Without the correct emissivity the object's temperature cannot be measured. Only differences in temperature can be obtained from a surface with the same emissivity. An example is shown in figure 4.3. It shows surfaces with different emissivity

FIGURE 4.3: Surfaces with different emissivity, left visible spectrum, right IR photograph. The surface temperatures are the same, temperature differences in the IR picture are caused by different emissivity of surfaces

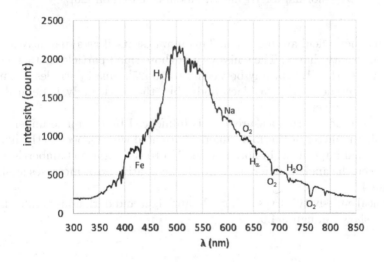

FIGURE 4.4: Measured solar spectrum

but with the same temperature. The different emissivity is caused by different surfaces. In the IR photo the surfaces therefore seem to have different temperatures.

Real objects are different from ideal black bodies. They do not radiate all the energy. An example is shown in figure 4.4.

Figure 4.4 also shown the absorption in the atmosphere at certain wavelengths, in this case especially by water vapor, oxygen and CO_2.

It is also important to note that emissivity is defined for macroscopic objects that are significantly larger then the wavelength of used radiation. When the object is of similar size as the wavelength it does not apply. This is for example the case of small particles such as dust or soot. They influence the

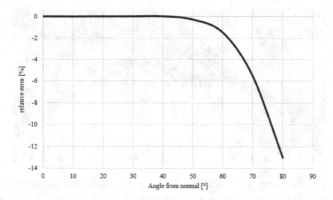

FIGURE 4.5: Typical shape of relative error as a function of angle from
normal for IR measurements—based on [26]

electromagnetic field around them. This manifest itself as apparently a much
larger emissivity than 1. The emissivity of those small particles is then significantly larger that 1, typically between 5 to 10 [25]. Small particles seemingly
radiate more energy that a black body, but the black body is defined as a
macroscopic object.

The accuracy of IR measurement is influenced by the angle under which
the IR thermometer or thermal camera is oriented with respect to the normal.
The maximal intensity is in normal direction to the surface (Lambert's cosine
law). When the angle is different from 90, the relative error increases as shown
in figure 4.5.

Typical emissivity curves for metals, and light and dark non-metallic materials are shown in figure 4.6.

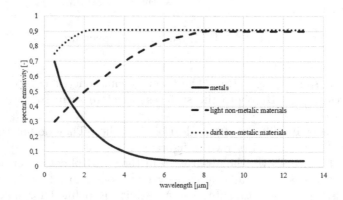

FIGURE 4.6: Typical emissivity curves for metals, and light and dark
non-metallic materials—based on [26]

4.1 Absorption in atmosphere and in materials

For non-contact temperature measurement, it is necessary to select the wavelength with the lowest attenuation. Based on the requested temperature range, a suitable atmospheric window has to be chosen. The smallest attenuation is achievable in the Short-Wavelength InfraRed (SWIR), Middle-Wavelength InfraRed (MWIR) and Long-Wavelength InfraRed (LWIR) windows. In other parts of the spectrum, the IR radiation is attenuated, most often by water vapors.

The attenuation is shown in figure 4.7. If the IR measurement would be done outside of those available windows, the attenuation would be very large. Not all materials transparent in the IR spectrum are transparent in visible spectrum as shown in figure 4.8.

4.2 Materials for IR optics

As apparent from above, non-contact temperature measurement operates almost exclusively in the IR spectrum. The IR optics will therefore require different materials from visible optics. Typically silicon or germanium are used. As shown in table 4.3 there are other materials available as well. Some of them are suitable for the visible and IR spectrum as well.

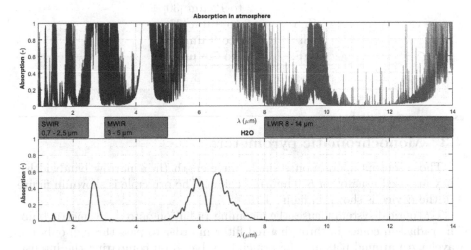

FIGURE 4.7: Top—total absorption in the atmosphere; data from [27]; Gemini Observatory, bottom—absorption in water vapors—data from [28]

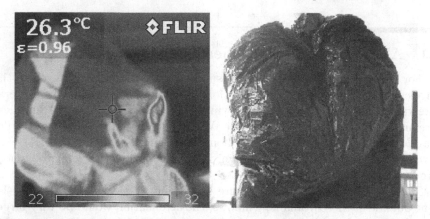

FIGURE 4.8: Absorption in material—not all materials transparent in the visible spectrum are transparent for IR and vice versa. Left: IR image; right: visible spectrum—photo author

TABLE 4.3: Materials for IR optics

Material	Usable range
BaF2	0,25 to 9.5 μm [29]
CdTe	1 to 25 μm [29]
CaF2	0,13 to 10 μm [29]
SiO2	0,25 to 3,5 μm [29]
GaAs	2 to 15 μm [29]
ZnSe	1 to 15 μm [30]
Ge	2 to 12 μm [30]
ZnS	8 to 8 μm [30]
Si	1,2 to 9 μm [31]
MgF2	0,11 to 7,5 μm [31]

4.3 Monochromatic pyrometers

Those sensors operate on a single wavelength (in a narrow bandwidth). They are used usually for 650 nm (red color). The principle is shown in figure 4.9; the device is shown in figure 4.10.

The optical system focuses the incoming radiation onto a pyrometric bulb. The radiation comes in through a red filter in order to pass through only the wavelength around 650nm. The principle is based on comparing the intensities of the incoming radiation with the intensity of the pyrometric bulb on a single wavelength. The intensity of the pyrometric bulb's filament is set

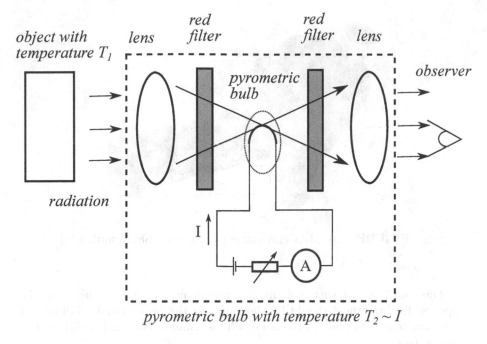

object with temperature T_1 — lens — red filter — red filter — lens — observer

pyrometric bulb

radiation

I

A

pyrometric bulb with temperature $T_2 \sim I$

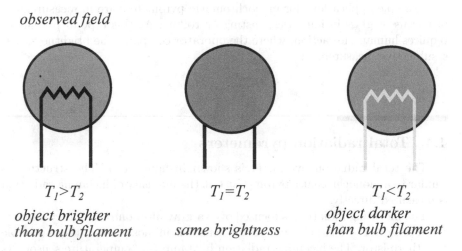

observed field

$T_1 > T_2$	$T_1 = T_2$	$T_1 < T_2$
object brighter than bulb filament	*same brightness*	*object darker than bulb filament*

FIGURE 4.9: Principle of monochromatic pyrometer

with current. If the intensities are the same, the temperatures are the same as well.

If the filament's temperature is lower, we see a dark filament on a bright background. If the filament's temperature is higher, we see a bright filament on a darker background.

FIGURE 4.10: Monochromatic pyrometer—photo author

The goal is to set such a filament temperature so that the filament disappears. Its brightness should be the same as the background's brightness. The current is then read on the scale, where filament current is calibrated as temperature.

Example applications for monochromatic pyrometers are in measurement of metals or glass in foundries, casting or rolling. As this type of pyrometer requires human interaction, where the operator compares the brightness, it is a subjective measurement.

4.4   Total radiation pyrometers

The total radiation pyrometer is shown in figure 4.11. The structure is similar to a monochromatic pyrometer, but the intensity of incoming radiation is measured directly.

The incoming radiation is focused over a gray filter onto a contact temperature sensor. The contact temperature sensor is in most cases a thermocouple or a thermistor. The incoming radiation heats up the temperature sensor. Its temperature is then shown directly on a scale.

Total radiation pyrometers are usable in ranges between approximately −40 to +5000°C based on the used detector. They are used to measure temperature in furnaces for example. Example devices and applications are shown in figure 4.12 and figure 4.13

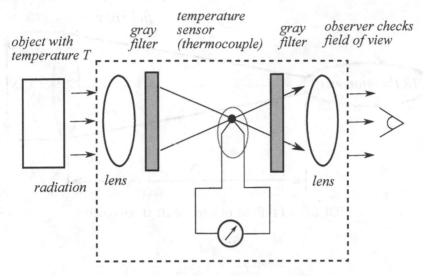

FIGURE 4.11: Total radiation pyrometer

FIGURE 4.12: Total radiation pyrometer—courtesy of [32], with permissions from Optron GmbH, Germany

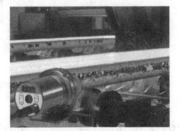

FIGURE 4.13: High-optical resolution two-colour pyrometer "Cella Temp PX 40" for measuring from a safe distance. courtesy of KELLER HCW GmbH, Germany [33] (background picture: © Can Stock Photo/jordache, with permissions from KELLER HCW GmbH, Germany

4.5 IR thermometers

IR thermometers are in principle total radiation pyrometers. They use a thermocouple or a bolometer as sensor.

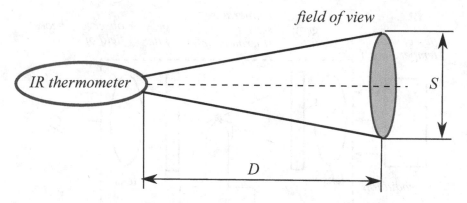

FIGURE 4.14: Field of view of IR thermometer

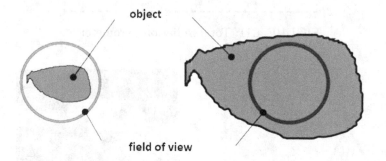

FIGURE 4.15: Field of view of an IR thermometer, right correct field of view
size vs. object size, left incorrect field of view size vs. object size

For all IR measurements the correct use is very important. Every IR thermometer has a given field of view. It does not measure in a single point but in
an area as shown in figure 4.14. For a correct measurement it is important that
the field of view is smaller than the measured object. Then only the object's
radiation and not background radiation is measured. See figure 4.15. If the
measured object is of similar size as the field of view, the measurement can
be correct as well, but care needs to be taken about the temperature on the
object's edges that can typically be lower.

The incorrect measurement is done when the measured object is smaller
than the field of view. In this case the sum of objects' radiation and background
radiation is measured. The radiation intensity does not represent correctly the
object's temperature.

The size of the field of view is usually given with a ratio between its diameter and distance. For example 6:1. This means that for a distance between
the IR thermometer of 6 cm the field of view has a diameter of 1 cm.

4.6 Principles of correct use of IR thermometers

In order to measure correctly with an IR thermometer or thermal camera, consider the basic fact: **Intensity is measured not temperature.** If we want to measure temperature, we need to know emissivity and limit the influence of other IR sources.

An example of two objects with a different emissivity is shown in figure 4.16. They seem to have different temperatures, but the temperature is in fact the same. The difference is caused by different emissivity (metal, wood,...). The same temperature was verified with a contact thermometer.

An example of an object with low emissivity is shown in figure 4.17. Reflections are visible. Unless the reflections are eliminated, the object's surface temperature can't be measured. The object shown in the figure is a mirror.

In order to "set" known emissivity on the surface, stickers shown in figure 4.18 or paint shown in figure 4.19 are used.

An example showing the importance of correct emissivity setting is shown in figure 4.20. The figure shows a heated plate. The plate is painted with different colors; they have different emissivities. Although the temperature can't change so rapidly on the thermograph it looks like it. The apparent temperature changes are caused by the different emissivity of the surfaces. For example in the top-right corner there is a sharp transition between the silver and black area. The heat source is placed behind the black area. The dark strip (seemingly colder area) is an aluminium tape with low emissivity that reflects IR radiation. It looks like a cold spot, but in fact has the same temperature as the other areas.

FIGURE 4.16: Different temperature reading caused by different emissivity of the surfaces—photo author

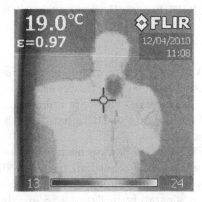

FIGURE 4.17: Surface with low emissivity, reflection from the person in front of a mirror—photo author

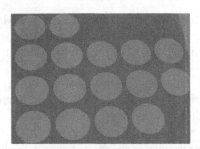

FIGURE 4.18: Stickers with defined emissivity for IR measurement—photo author

FIGURE 4.19: Paint with defined emissivity for IR measurement—photo author

FIGURE 4.20: Heated surface with different emissivities, left IR, right visible spectrum—photo author

Also the stickers with defined emissivity are visible. They have a defined emissivity of 0.96. The temperature can be read correctly only on those stickers if the other emissivities are unknown.

The blocking of IR radiation with a piece of ordinary glass is shown in figure 4.21.

Figure 4.22 shows a heatsink. One half (left) is black painted with emissivity 0.96. The other half (right) is shiny aluminum. Both surfaces seem to have a different temperature although it is in fact the same. Also the stickers, shown separately in figure 4.18 are shown, with a defined emissivity of 0.96 are visible. They can be used to measure the correct temperature.

The use of paint with defined emissivity is shown in figure 4.23 where the temperatures of Cu tubes is shown.

FIGURE 4.21: Glass with reflections—photo author

FIGURE 4.22: Heated heatsink, every half with different emissivity, temperature the same. Stickers with emissivity 0.96 apparent—photo author

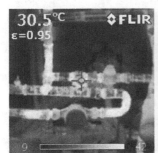  

FIGURE 4.23: Temperatures of Cu tubes—"hot" = bright spots are painted with paint with known emissivity of 0.96, temperature can be read correctly only here—photo author

It is obvious that without the knowledge of emissivity it is not possible to measure the absolute temperature. Only temperature differences on a surface with constant emissivity can be determined.

5

Force

The most used force sensor is a strain gauge. The force is not measured directly but by means of deformation. Deformation changes the electrical resistance of the strain gauge, which is then measured.

Strain gauges can be made from both metals and semiconductors. The electric resistance is dependent on the material (described by resistivity), length and cross section.

When a force is applied on the strain gauge, it changes its dimensions and hence the electrical resistance.

5.1 Metallic strain gauges

The electric resistance of a wire is

$$R = \rho \frac{l}{A} \tag{5.1}$$

where ρ is resistivity, l is wire length and A is wire cross section.

The real foil strain gauge is shown in figure 5.1, the structure of a strain gauge is shown in figure 5.2.

The measurement grid of a strain gauge is produced from a thin wire (a wire strain gauge) or from a thin foil (foil strain gauge).

In both cases, the wire or foil must be produced from a material with a very small temperature coeffitient of electrical resistance.

The most common material is constantan (an alloy of Ni 45% and Cu 55%). Nevertheless it is necessary to use temperature compensation as will be shown later.

The production of strain gauges uses similar technologies as production of printed circuit boards (PCB) and etching of the laminated foil on a backing material. The backing material is most often paper or polyamid.

The change of electrical resistance is very small.

The electrical resistance is

$$R = \rho \frac{l}{A} \tag{5.2}$$

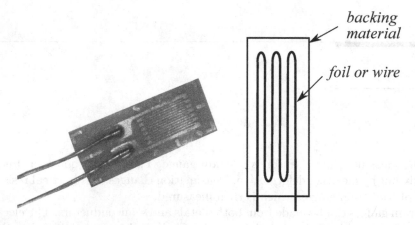

FIGURE 5.1: Foil strain   FIGURE 5.2: Structure of a strain
gauge—photo author   gauge

It can be seen that the change of electrical resistance depends on three parameters:
- change of electrical resistance with strain
- change of electrical resistance with change of length
- change of electrical resistance with change of cross section

For metal stain gauges the change of resistivity with strain is negligible, the strain gauge is changing its electrical resistance with changes of geometry.

The change of resistivity with strain (Piezo-resistive effect) is important for semiconductor strain gauges. For semiconductor strain gauges the changes of resistance caused by geometry are only about 2% [34]. The resistance change is caused by changes in electron mobility.

When a force is applied to an object, as shown in figure 5.3, the length increases.

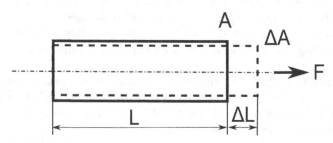

FIGURE 5.3: Change of dimensions with applied force

The change of electrical resistance is

$$\frac{dR}{R} = \frac{dl}{l} - \frac{dA}{A} + \frac{d\rho}{\rho} \qquad (5.3)$$

The dependence of length and cross section is described with the Poisson's ratio ν

$$\frac{dA}{A} = -2\nu\frac{dl}{l} \qquad (5.4)$$

The Poisson's ratio ν for common technical materials is in the range (0,2 to 0,5), for example, steel 0.3; aluminum 0.33; rubber 0.5.

The equation can be further simplified

$$\frac{dR}{R} = \frac{dl}{l} - \frac{dA}{A} + \frac{d\rho}{\rho} = (1+2\nu)\frac{dl}{l} + \frac{d\rho}{\rho} \qquad (5.5)$$

yielding

$$\frac{dR}{R} = K\frac{dl}{l} = \left[(1+2\nu) + \frac{\frac{d\rho}{\rho}}{\frac{dl}{l}}\right] \cdot \frac{dl}{l} = K \cdot \varepsilon \qquad (5.6)$$

Constant K is called the gauge factor.

The term $(1+2\nu)$ in the equation is expressing the changes of electrical resistance caused by "pure geometry." It is significant for metal strain gauges.

The term $\frac{\frac{d\rho}{\rho}}{\frac{dl}{l}}$ is caused by the piezoresistive effect and is significant for semiconductor strain gauges.

As obvious, the dependence of electrical resistance R and strain ϵ is linear for metal stain gauges.

$$\frac{dR}{R} = K \cdot \varepsilon \qquad (5.7)$$

For metal strain gauges the gauge factor is approx. 2.1. It is, of course, dependent on the material, but this is the value for constantan.

Typically strain $\epsilon = \frac{\Delta l}{l}$ is expressed in units $[\frac{m}{m}]$.

Table 5.1 shows the gauge factors for selected materials. Based on the numbers, it may seem that platinum might be a suitable material. However platinum has a relatively large temperature coefficient of resistance. Hence it is used in temperature sensors and not for strain gauges. Similar situation is for many other metals.

Nickel might seem also as a suitable material with a high sensitivity (−12.1). The sensitivity however changes with strain.

The typical basic resistance for a strain gauge is 120 Ω. Also 350 Ω is used. This resistance is specified for temperature of 23°C.

TABLE 5.1: Gauge factor for selected materials—based on data from [35]

Material	Gauge factor
Pt 100%	6.1
Pt 95%, Ir 5%	5.1
Pt 92%, W 8%	4.0
Isoelastic (Fe 55.5%, Ni 36% Cr 8%, Mn 0.5%)	3.6
Constantan / Advance / Copel (Ni 45%, Cu 55%)	2.1
Nichrome V (Ni 80%, Cr 20%)	2.1
Karma (Ni 74%, Cr 20%, Al 3%, Fe 3%)	2.0
Armour D (Fe 70%, Cr 20%, Al 10%)	2.0
Monel (Ni 67%, Cu 33%)	1.9
Manganin (Cu 84%, Mn 12%, Ni 4%)	0.47
Nickel (Ni 100%)	−12.1

5.1.1 The importance of temperature compensation

The importance of temperature compensation is shown on a calculation example. A steel bar with dimensions 10×10 mm, Young's modulus E = 200 GPa. The bar is loaded with a force $F = 1000$ N—see figure 5.4.

There are two strain gauges installed on the steel bar. One made from constantan, the other **fictive** made from copper.

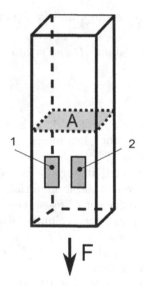

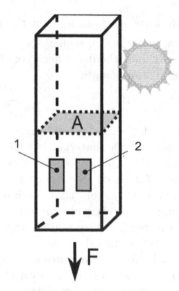

FIGURE 5.4: Experiment temperature compensation

FIGURE 5.5: Experiment temperature compensation—heated bar

It has to be stressed that copper is not used for real strain gauges, it is used here for demonstration only to see the effects of temperature changes.

The tension is

$$\sigma = \frac{F}{A} = \frac{1000\,[\mathrm{N}]}{1\,[\mathrm{cm}^2]} = 10\,[\mathrm{MPa}] = 0,01\,[\mathrm{GPa}] \tag{5.8}$$

Strain is

$$\varepsilon = \frac{\sigma}{E} = \frac{0,01\,[\mathrm{GPa}]}{200\,[\mathrm{GPa}]} = 50 \cdot 10^{-6} \tag{5.9}$$

The elongation is very small, only $50 \cdot 10^{-6}$. Usually the elongation is in the range of

$$\varepsilon = 10^{-5} \div 10^{-3}.$$

For simplification the calculation will consider the same gauge factor for both strain gauges. K = 2 will be considered (In reality copper has K = 2.6). The change of electrical resistance is

$$\frac{dR}{R} = K \cdot \varepsilon = 2 \cdot 50 \cdot 10^{-6} = 0,1 \cdot 10^{-3} \tag{5.10}$$

For a strain gauge with $R = 120\ \Omega$ the absolute change of electrical resistance is

$$dR = K \cdot \epsilon \cdot R = 2 \cdot 50 \cdot 10^{-6} \cdot 120 = 0,1 \cdot 10^{-3} \cdot 120 = 12\ \mathrm{m\Omega} \tag{5.11}$$

As a circuit a quarter-bridge will be used. The power supply voltage is 12 V. The output voltage is

$$V_0 = V_s \cdot \frac{K}{4} \cdot \varepsilon = 12 \cdot \frac{2}{4} \cdot 50 \cdot 10^{-6} = 0,3\,[\mathrm{mV}] \tag{5.12}$$

The bridge output voltage is very small, the circuit will require a DC amplifier.

Now let's heat the steel bar by just 10°C—see figure 5.5.

The change of electrical resistance caused by the temperature change for the constantan strain gauge is

$$R(T) = R(T_0)\,(1 + \alpha\Delta T) = 120\,(1 + 8 \cdot 10^{-6} \cdot 10) = 120,0096\,[\Omega] \tag{5.13}$$

$$\Delta R(T) = 9,6\,[\mathrm{m\Omega}] \tag{5.14}$$

The change of electrical resistance caused by the temperature change for the copper strain gauge is

$$R(T) = R(T_0)\,(1 + \alpha\Delta T) = 120\,(1 + 3,9 \cdot 10^{-3} \cdot 10) = 124,68\,[\Omega] \tag{5.15}$$

$$\Delta R(T) = 4,68\,[\Omega] \tag{5.16}$$

It is obvious that the change of electrical resistance caused by a temperature change of only 10°C is very large. For the fictive copper strain gauge it is about 500× larger that for the constantan one.

The calculation showed that a temperature compensation is absolutely essential. Without temperature compensation the strain gauge measures temperature changes and not strain. Only after the temperature is compensated, will the strain gauge measures strain.

The changes of electrical resistance are also very small. The changes are approximately tenths to hundreds of mΩ. The small changes are not measured directly but with special methods described in next chapters.

5.1.2 Strain-gauge rosettes

As obvious from previous explanation, the strain gauge is sensitive to changes of dimensions. In the ideal case it should be sensitive only to one direction of force and not the lateral one as shown in figure 5.6.

In order to measure strain, forces or deformation in multiple axes, straingauge rosettes are used. The rosette uses two or three strain gauges on the same backing material installed under certain angles. Multiple strain gauges can be used as well. An example of a strain-gauge rosette is shown in figure 5.7.

If the direction of main stresses are known, a strain-gauge rosette with two strain gauges can be used. The strain gauges are installed under the angle of 90°.

For applications where the direction of main stresses are unknown, straingauge rosettes with angles 0°, 60°, 120° are used.

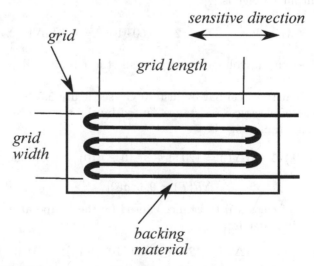

FIGURE 5.6: Main components of a foil strain gauge

FIGURE 5.7: Strain-gauge rosette 120° to 120 Ω—photo author

The main stress is calculated as follows [34]

$$\sigma_{1,2} = \frac{E}{1-\nu} \cdot \frac{\epsilon_a + \epsilon_b + \epsilon_c}{3} \pm \frac{E}{1+\nu} \cdot \sqrt{(\frac{2\epsilon_a - \epsilon_b - \epsilon_c}{3})^2 + \frac{1}{3}(\epsilon_b - \epsilon_c)^2} \quad (5.17)$$

The temperature range for constantan strain gauges is approximately −200°C to +200°C, in sensors they are usually used in a temperature range approximately −20°C to +70°C [34].

5.2 Semiconductor strain gauges

For some applications, semiconductor stain gauges can have some advantages. It especially concerns spatially restricted installations where metal strain gauges might be too large. Examples of semiconductor strain gauges without backing are shown in figure 5.8. They are also usable in applications where the temperature is constant because semiconductor strain gauges have a large temperature dependence.

Compared to metal, semiconductor strain gauges have significantly larger sensitivity. The gauge factor is > 100, it is about 25× larger that for metal strain gauges.

The gauge factor for a p-type semiconductor strain gauge is approximately +110 to +130; for n-type semiconductor strain gauges, the gauge factor is approximately −80 to −100 [34].

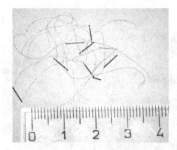

FIGURE 5.8: Example of semiconductor strain gauges—photo author

The deformation sensitivity is

$$K = 1 + 2\mu + \pi_i \cdot E_d \tag{5.18}$$

where π_i is the piezoresistive coefficient in the longitudinal direction, and E_d is the Young's modulus in tension.

Semiconductor strain gauges have a significant temperature dependence [36]

$$R_{0,t} = R_{0,25}(1 + a(t - 25) + b(t - 25)^2) \tag{5.19}$$

where a and b are temperature coefficients of uninstalled strain gauge and $R_{0,25}$ is the resistance for 25°C [Ω].

Compared to metal strain gauges, the semiconductor strain gauges are about 75× more sensitive to temperature.

Also the dependence of resistance on strain is non-linear [36] for semiconductor strain gauges.

$$R_{\epsilon,25} = R_{0,25}(1 + C_1 \cdot \epsilon + C_2 \cdot \epsilon^2) \tag{5.20}$$

where C_1, C_2 are coefficients of the equations.

5.3 Circuits for strain gauges

In order to measure the small changes of resistance, special circuits have to be used. Bridge methods will amplify the small resistance change and they will also provide temperature compensation if connected correctly.

The bridge power supply voltage can be AC or DC. Commercial bridges work as voltage supplies with voltage about 1 to 10 V. The voltage is variable and is set in function of the strain-gauge current.

The bridge output is then usually connected to an amplifier that amplifies its small voltage.

The disadvantage is that a DC-supplied bridge can be influenced with electric or magnetic fields or with thermoelectric voltages. Those small voltages are amplified in the signal amplifier as well.

In case of a AC power supply, only the AC signal is amplified and the interference signals are suppressed. The commonly used frequencies are 225 Hz and 5 kHz. The former frequency is used for static and quazi-static measurements up to approximately 9 Hz. The later frequency is used for both static and dynamic measurements up to about 1 kHz [34].

Quarter-bridge This connection contains one measurement strain gauge. This strain gauge measures the measured deformation. Another strain gauge is used for temperature compensation. The connection is shown in figure 5.9.

All connections shown here include temperature compensation. In all following schematics, the elements for temperature compensation are marked as R_C.

All resistances will follow the notation shown in figure 5.10. All following schematics will use the same notation and placement for $R_1, ..., R_4$.

The used strain gauge has a resistance of R_0, for example 120 Ω. The resistance is changed in range $\pm\Delta R$.

From the condition of balance it follows

$$R_1 \cdot R_4 = R_2 \cdot R_3 \tag{5.21}$$

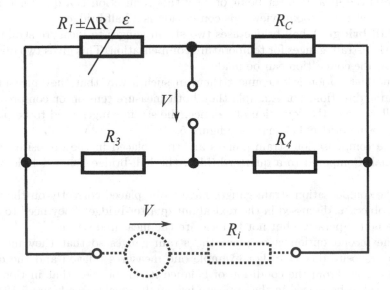

FIGURE 5.9: Strain gauge—quarter-bridge, R_1 = sensor, R_C = temperature compensation

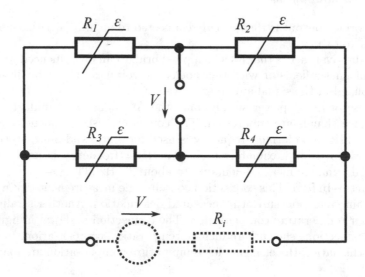

FIGURE 5.10: Notation of elements in a bridge

that the compensating strain gauge has to be placed in the opposite diagonal than the measurement strain gauge. The compensating strain gauge also needs to have the same temperature as the measurement one. This has to be assured with correct placement as will be shown later.

From the measurement point of view this connection is a quarter-bridge; from the electrical point view this connection is a half-bridge.

Half-bridge A half-bridge uses two strain gauges to measure strain and two other strain gauges for temperature compensation. This gives two options for how the connection can be made.

The first option is to connect them in such a way that they measure in the same direction. For example they both measure tension or compression. As follows from the condition of balance, the strain gauges need to be placed on the same bridge diagonal—see figure 5.11.

The compensation strain gauges are then placed in the opposite bridge diagonal. Compared to a quarter-bridge, the half-bridge has a double sensitivity.

The compensation strain gauges have to be placed correctly on the measured object as discussed in the text about quarter-bridge. They need to have the same temperature but not to measure the measured strain.

The second option is to place the strain gauges so that they measure in the opposite direction. For example one measures tension and the other compression. From the condition of balance it can be seen that in this case they need to be placed in the opposite bridge diagonals—see figure 5.12.

Full bridge A full bridge uses all four strain gauges to measure strain. The same strain gauges are also used for temperature compensation. Two have to measure tension, and the two others have to measure compression. They

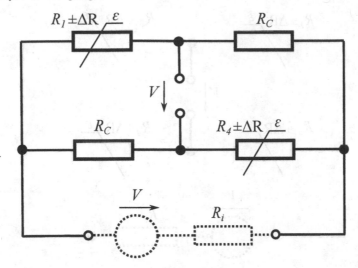

FIGURE 5.11: Half-bridge—sensors measure in the same direction, R_1, R_4 = sensor, R_C = temperature compensation

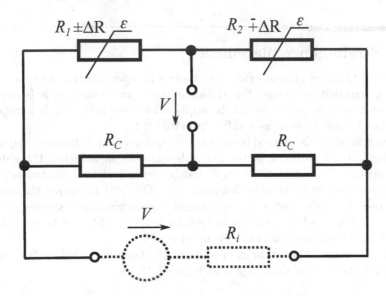

FIGURE 5.12: Half-bridge—sensors measure in the opposite direction, R_1, R_4 = sensor, R_C = temperature compensation

also need to be installed in such as way that they have the same temperature. Under this condition, the connection is automatically self-compensating for temperature changes. The connection is shown in figure 5.13.

The full bridge doubles the sensitivity as compared to a half-bridge.

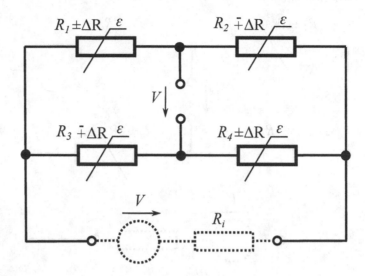

FIGURE 5.13: Full bridge

5.4 Strain-gauge placement

The installation of strain gauges is made with specialized glues supplied by strain gauge manufacturers. Based on theirs recommendations it is required to select a suitable glue type for the application—see table 5.2. It is required that the glue thickness is as small as possible [34].

The grid length is selected based on the application and the measured material. One of the factors can be the available space for installation. If the installation space is limited, then the strain gauge type selection is limited as well.

Another factor is material homogeneity. The grid measures the average strain over its whole length. If the material is homogeneous a shorter grid can be chosen. For non-homogeneous material a longer grid has to be used. It has to be longer than the non-homogeneity size.

The following installation examples are all shown for a full-bridge connection, with the resistor markings from figure 5.10.

TABLE 5.2: Glues for strain gauges—according to [34]

glue	E [kN/mm^2] (0 .. 50°C)
X 60 (acrylic)	4,5 .. 6
Z 70 (cyanoacrylic)	3
H (polyester)	3,5
EP 250, EP 310 (epoxy)	2,8
CR 760 (ceramic paste)	20 .. 21
constantan	163

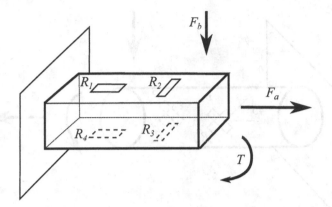

FIGURE 5.14: Strain gauge placement to measure axial force

As the full bridge has the highest sensitivity and automatic temperature compensation it is the most common connection.

The placement to measure the axial force with bending, torque and temperature compensation is shown in figure 5.14. It uses strain gauges R_1 and R_4 to measure the axial force. Gauges R_2 and R_3 compensate for torque T and bending force F_b.

All strain gauges together then compensate for eventual temperature changes. The assumption is that they all have the same temperature.

The placement to measure the bending force with axial force and temperature compensation is shown in figure 5.15. It uses strain gauges R_1, R_4, R_2 and R_3 to measure and compensate the axial force. All strain gauges are stretched equally by the axial force. When downward bending force is applied,

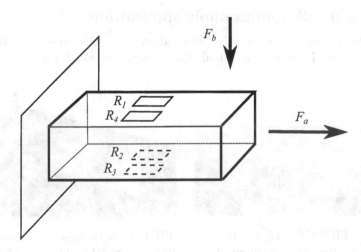

FIGURE 5.15: Strain gauge placement to measure bending force

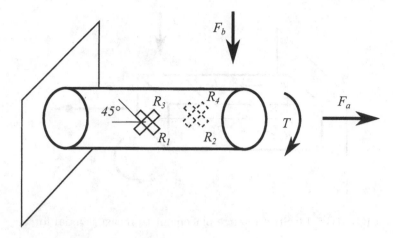

FIGURE 5.16: Strain gauge placement to measure torque

strain gauges R_1 and R_4 measure tension; strain gauges R_2 and R_3 measure compression. All strain gauges together compensate for temperature.

In order to measure torque, the strain gauges are place under the angle of 45° along the axis. This is because this is the direction of main stresses—see figure 5.16.

A strain-gauge rosette can be used, with two strain gauges under the angle of 90°.

5.5   Load cells and example applications

Strain gauges can be installed either directly on the measured object or on a load cell. Load cells are used, for example, in weighting applications.

FIGURE 5.17: Strain measurement on composite material

FIGURE 5.18: Beam-type load cell in a weight—photo author

FIGURE 5.19: Type-S load cell—photo author

FIGURE 5.20: Measurement of residual stress after heat treatment—courtesy of [37], with permissions from Stresscraft Ltd, UK

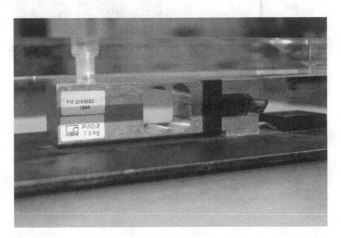

FIGURE 5.21: Beam-type load cell—photo author

The load cell is being deformed with the applied external force. An example of a strain gauge placed on a composite material is shown in figure 5.17. A beam-type load cell is shown in figure 5.18; an S-type load cell is shown in figure 5.19. A strain-gauge rosette used for residual stress measurement is shown in figure 5.20. A beam load cell in weighting application is shown in figure 5.21. The principle of an S-type load cell is shown in figure 5.22, the load cell itself is then shown in figure 5.23, where it was used on a torque dynamometer.

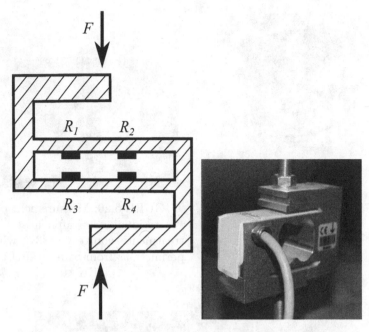

FIGURE 5.22: Type-S load cell

FIGURE 5.23: Type-S load cell—photo author

6

Torque

6.1 Torque dynamometer

A torque dynamometer is in principle a rotating electric machine. It can be an AC or DC machine.

The principle of action and reaction is used, as shown in figure 6.1. The machines stator is allowed to rotate. The torque from the shaft is transferred through the air gap to the stator. The stator has the tendency to rotate. This is prevented for example with a spring. The stator's rotation is proportional to the measured torque. A torque dynamometer is shown in figure 6.2.

Typically the force on a known lever is measured with a force sensor. Also a weight can be used. In principle, any strain gauge described in the previous chapter can be used.

The torque is calculated as

$$T = F \cdot l \tag{6.1}$$

where F is the measured force and l is the used lever length.

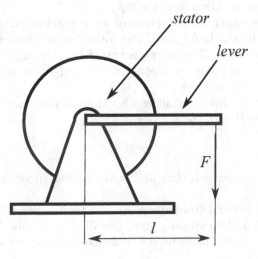

FIGURE 6.1: Principle of torque dynamometer

FIGURE 6.2: Torque dynamometer—force sensor on the left, S-type load
cell, speed sensor on the left top (tachodynamo), photo author

6.2   Electronic torque sensors

This type of sensor uses strain gauges on a shaft. The strain gauges are
installed under the angle of 45° to the shaft axis. A full bridge connection is
used as it has the highest sensitivity and automatic temperature compensation. The principle is shown in figure 6.3.

The bridge is either battery-powered or a wireless (transformer) power
supply is used. In the latter case, the rotor has a winding, and a second
winding is on the stator. The energy is transferred through the air gap.

In modern sensors the signal is converted to a digital signal and transferred
by radio waves to the receiver.

Some sensors of this type allow also the calculation of the transferred
mechanical power P with the use of equation

$$P = T \cdot \omega \tag{6.2}$$

where T is the measured torque and ω is angular speed, measured with a
speed sensor.

Examples of electronic torque sensors and their applications are shown in
figures 6.5 and 6.6. Two strain gauges installed on a tube for static torque
measurement are shown in figure 6.4. Industrial torque sensors are shown in
figures 6.7 and 6.8.

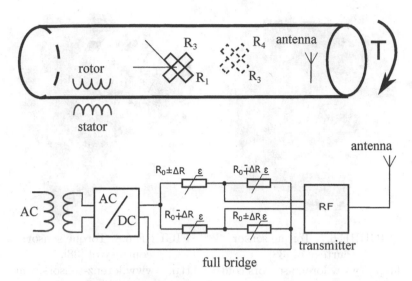

FIGURE 6.3: Shaft torque sensor with wireless transfer

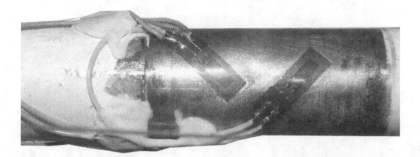

FIGURE 6.4: Torque measurement on a fixed (not rotating) pipe—photo author

FIGURE 6.5: Torque sensor—photo author

FIGURE 6.6: Torque sensor for tram wheel testing—photo author

FIGURE 6.7: Torque sensor, courtesy of [38], http://www.lorenz-sensors.com with permissions from Lorenz Messtechnik GmbH, Germany

FIGURE 6.8: Torque sensor, courtesy of [38], http://www.lorenz-sensors.com with permissions from Lorenz Messtechnik GmbH, Germany

7

Position

Position sensors measure either the position of an object or they detect its presence. They can be divided into several categories:

Relative or absolute:

Absolute—absolute position sensors send out the correct position right after they are powered. No movement to a reference position is required. They have usually lower resolution than relative sensors.

Relative—relative sensors don't give the correct reading after power up. In most cases they count pulses from a reference position. A movement to a reference (zero) position is required. The pulses are then counted from this reference position. The step size for one pulse is known. Compared to absolute sensors they usually have a much higher resolution.

Analog or digital:

This is related to the sensor principle and not to the sensor output signal.

Analog—analog position sensors are continuous. They have in theory an infinite resolution. The resolution is limited by the subsequent circuits that process the signal (such as A/D converters, etc.).

Digital—digital position sensors are non-continuous. They have a defined minimal resolution. The digital signal is then most frequently transferred as a digital signal, although it could be converted to an analog signal as well (current or voltage).

Continuous or proximity:

Continuous—they measure continuously the distance between the sensor and the object. The output signal is proportional to the measured distance.

Proximity—proximity sensors detect only the presence of the object not its distance. The output is a digital signal, 0/1, object is present or not. Proximity sensors are used for example on production or packaging lines to detect the products. The distance is not important, the object is where it should be, and it can be processed or packed. Proximity sensors are less expensive than continuous sensors.

Linear or angular:

Linear—the sensor measures in principle linear displacement

Angular—the sensor measures in principle angular displacement

Gears can be used to transfer from linear to angular motion and vice versa. However it is typically more convenient to select a sensor for the measured type of displacement than to have trouble with gears.

7.1 Resistive sensor

In principle this is a variable resistor, as shown in the schematic in figure 7.1. The slider is connected to the measured object. As the position changes, the resistance changes, and the output voltage is a function of position.

The shown connection is a loaded voltage divider. Its output is loaded with resistance R_L. The output voltage is

$$V_2 = \frac{\frac{R_L R_2}{R_L + R_2}}{\frac{R_L R_2}{R_L + R_2} + R_1} \cdot V = V \frac{R_L R_2}{R_1 R_2 + R_1 R_L + R_2 R_L} \tag{7.1}$$

The slider position x changes the electrical resistance R_2

$$R_2 = x R_0 \tag{7.2}$$

$$R_1 = R_0 - R_2 = R_0 - x R_0 = (1 - x) R_0 \tag{7.3}$$

$$k = \frac{R_L}{R_0} \tag{7.4}$$

The sensor's output voltage therefore is

$$\begin{aligned} V_2 &= V \frac{x \cdot k \cdot R_0 \cdot R_0}{(1-x) R_0 \cdot x \cdot R_0 + (1-x) R_0 \cdot k \cdot R_0 + x \cdot k \cdot R_0 R_0} = \\ &= V \frac{x \cdot k \cdot R_0^2}{x \cdot R_0^2 - x^2 \cdot R_0^2 + x \cdot R_0^2 - x \cdot k \cdot R_0^2 + x \cdot k \cdot R_0^2} = \\ &= V \frac{k \cdot x}{(1-x) x + k} \end{aligned} \tag{7.5}$$

The shown equation is non-linear for $k < \infty$, i.e., for a load resistance $R_L < \infty$.

It is, however, obvious that for the increasing load resistance R_L the non-linearity is getting smaller.

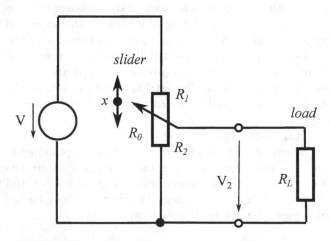

FIGURE 7.1: Schematic of resistive position sensor

For this type of sensor it is therefore required to select an instrument with a high input resistance. For instance a digital voltmeter. The typical input resistance of a digital voltmeter is at least 10 MΩ. When an analog voltmeter is used, its input resistance may be a lot smaller. The voltmeter is then loading the sensor with current and the non-linearity increases. This has then an effect on the sensor characteristic. The characteristic becomes significantly non-linear.

7.1.1 Relative error of resistive position sensor

The relative error is the relative difference between the real measured value (for $k < \infty$; load resistance $R_L < \infty$) and the ideal "correct" value of the output voltage (for $k- > \infty$; load resistance $R_L -> \infty$).

$$
\delta\,[\%] = \frac{V_2 - x \cdot V}{V} \cdot 100 = \frac{V\frac{k \cdot x}{(1-x)x+k} - x \cdot V}{V} \cdot 100 =
$$
$$
= -\frac{(1-x)x^2}{(1+x)x+k} \cdot 100 \tag{7.6}
$$

From the equation it is evident that the relative error in this case is always negative. It is zero for $k- > \infty$.

The measured voltage is always smaller than the ideal "correct" voltage that would be measured with an infinite voltmeter resistance. The relative error is increasing with decreasing load resistance.

The position of maximal relative error is

$$
\frac{\partial \delta}{\partial x} = 0 \Rightarrow x = 1/2 \tag{7.7}
$$

The position of maximal relative error is independent on the value of k. The steady-state response for a resistive position sensor with various values of K is shown in figure 7.2.

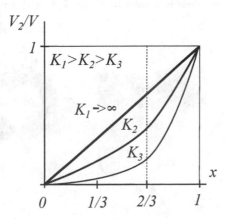

FIGURE 7.2: Steady-state characteristic of a resistive position sensor and its relative error

7.1.2   Absolute error

The position of maximal absolute error is

$$\Delta[V] = V\frac{k \cdot x}{(1-x)x + k} - x \cdot V \tag{7.8}$$

The position of maximal absolute error is dependent on the value of k. For example for $k = 1$ the position of maximal absolute error is

$$\frac{\partial \Delta}{\partial x} = 0 \rightarrow x \approx 0,69 \tag{7.9}$$

The position of maximal absolute error is approximately in 2/3 of the full range as shown in figure 7.4, where the relative error is shown as well.

7.1.3   Types of resistive position sensors

Wire The sensor is a wire-wound resistor. It has a shape of a coil. Those sensors have lower resolution than other types, but they are robust. The resolution is typically about 0.1% of full range. A resistive wire used in resistive sensor construction is shown in figure 7.3.

Metalized They have higher resolution then wire-wound resistors, the resolution is limited by the layer homogeneity. Possible problems with noise and wear may occur.

Carbon Wear sensitive material, used for low-cost variable resistors, not suitable for robust sensors

Conductive plastics Material used in industrial position sensors

CERMET (composite material—ceramics and metal) high-temperature resistance, linearity typically 0,05 – 0,1% (0,002% for larger diameters)

Examples of industrial resistive position sensors are shown in figure 7.5, where a sensor with conductive plastic is shown. A resistive position sensor used in a welding press is shown in figure 7.6. A draw-wire sensor is shown in figure 7.7.

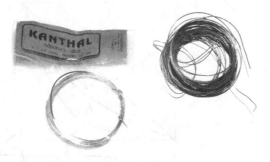

FIGURE 7.3: Resistive wire—photo author

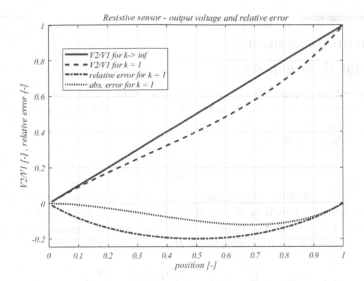

FIGURE 7.4: Absolute and relative error for a resistive position sensor

FIGURE 7.5: Resistive position sensor—photo author

FIGURE 7.6: Resistive position sensors on a welding press—photo author

FIGURE 7.7: Draw-wire resistive position sensor—photo author

7.2 Inductive sensors

7.2.1 Small air gap sensor

The small air gap sensor is used to measure small distances, up to approximately 1 mm. It is based on inductance L changes as the air gap length δ is changing. This is shown in figure 7.8.

The coil inductivity is

$$L = \frac{N^2}{R_m} \qquad (7.10)$$

where R_m is the magnetic resistance (reluctance) and N is number of turns.

Inductance as a function of air-gap size is shown in figure 7.9.

The reluctance in this magnetic circuit has two components. The air-gap reluctance R_{ag} and the iron reluctance R_{Fe}.

$$R_m = R_{Fe} + R_{ag} = \left(\frac{l_{Fe}}{\mu A}\right) + \left(\frac{2\delta}{\mu A}\right) \qquad (7.11)$$

where l_{Fe} is the length of iron magnetic circuit.

Assuming $R_{ag} \gg R_{Fe}$ then

$$L = \frac{\mu \cdot A \cdot N^2}{2 \cdot \delta} \qquad (7.12)$$

The changes of inductance are not evaluated directly, but the impedance change causes the change of inductor current. Current is then measured.

The coil impedance is

$$\hat{Z} = R + j\omega L \qquad (7.13)$$

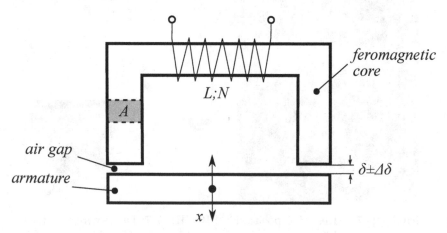

FIGURE 7.8: Small air gap sensor

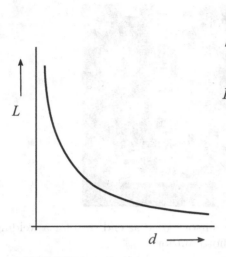

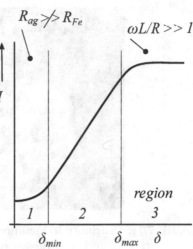

FIGURE 7.9: Inductance as a
function of air-gap size

FIGURE 7.10: Inductor current I
as a function of air-gap size

$$\hat{I} = \frac{\hat{V}}{\hat{Z}} = \frac{\hat{V}}{R + j\omega L} \tag{7.14}$$

Assuming

$$\frac{\omega L}{R} \gg 1 \tag{7.15}$$

The amplitude of current is

$$I = \frac{V}{\omega L} = \frac{2V}{N^2 \mu_0 A} \cdot \delta \tag{7.16}$$

In general this dependence is non-linear. Three distinct regions can be seen in the steady-state characteristic shown in figure 7.10.

Region 1 The assumption $R_{ag} \gg R_{Fe}$ is not valid, the air gap reluctance is of similar size to the iron reluctance. The air gap is small; the dependence is non-linear.

Region 2 The shown assumptions are valid, the air gap is sufficiently large, but not too large. The dependence is approximately linear. The sensor is used in this region.

Region 3 The assumption $\omega L/R \gg 1$ is not valid; the air gap is too large. Increase in the air gap does not change the current anymore.

Examples of industrial induction sensors and their applications are shown in figures 7.11, 7.12, 7.13.

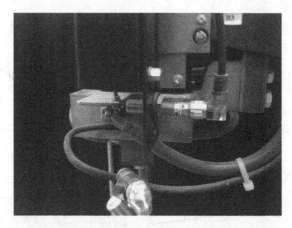

FIGURE 7.11: Proximity inductance sensor—used as end stop for a welding
press—photo author

FIGURE 7.12: Inductance   FIGURE 7.13: Continuous
sensor—photo author   inductive sensor—photo author

7.2.2   Linear variable differential transformer (LVDT)

This sensor is quite frequently used for both small and large ranges. It
has a very good accuracy. It is a robust sensor; the only moving part is a
ferromagnetic core. The movement of the core changes the mutual inductance
between the primary and secondary windings. The principle is shown in figure
7.14.

The sensor is powered with AC voltage into the primary winding. As the
core moves, the mutual inductances M1 and M2 between the primary and
secondary windings are changing. The voltage on the secondary windings can
be measured directly. The windings may also be connected in series as shown
in figure 7.14.

The steady-state characteristic is non-linear with a typical V shape as
shown in figure 7.15.

The non-zero voltage in the center ($x = 0$) is caused by stray magnetic
fluxes and capacitive coupling.

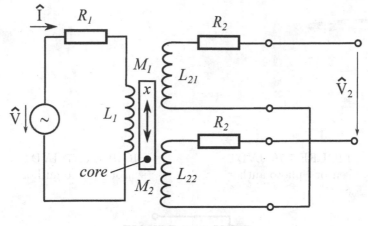

FIGURE 7.14: LVDT

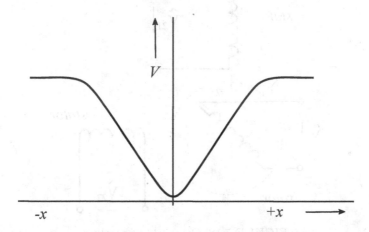

FIGURE 7.15: Steady-state characteristic of LVDT

LVDT sensors are robust and reliable, with low friction. They can be equipped with a spring that pushes the sensor tip against the object. Examples of industrial LVDT sensors are shown in figures 7.16 and 7.17.

7.2.3 Resolver

The resolver is in principle an electric machine. Is has one primary and two secondary windings as shown in figure 7.18. The secondary windings are shifted by 90°. It is an absolute sensor within the range of one revolution. The resolver is typically used as a sensor in electric machine control as shown in figure 7.19.

FIGURE 7.16: LVDT   FIGURE 7.17: LVDT
sensor—photo author   sensor—photo author

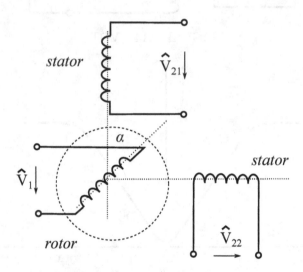

FIGURE 7.18: Principle of resolver

The rotor is AC powered with a voltage with frequency in the range approximately 2 to 20 kHz

$$\hat{V} = V_{\max} \sin(\omega t) \tag{7.17}$$

The induced voltage in stator winding is

$$\hat{V}_{21} = V_{\max} \sin(\omega t) \cdot \sin(\alpha) \tag{7.18}$$

$$\hat{V}_{22} = V_{\max} \sin(\omega t) \cdot \cos(\alpha) \tag{7.19}$$

The induced voltage is a function of the angle of rotation α. The dependence is sinusoidal for one and cosinusoidal for the other secondary winding. In order to get the absolute position in range 0° to 360° both voltages have to be measured.

FIGURE 7.19: Resolver—used to measure the angular position of a permanent magnet synchronous motor—photo author

FIGURE 7.20: Resolver to digital converter—photo author

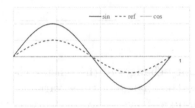

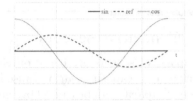

FIGURE 7.21: Resolver signal in position 0°, signal SIN with maximal amplitude, signal COS with amplitude zero

FIGURE 7.22: Resolver signal in position 90°, signal COS with maximal amplitude, signal SIN with amplitude zero

Specialized resolver-to-digital circuits exists, such as AD2S1200. An example of such a resolver-to-digital unit is shown in figure 7.20. The circuit provides all the required functions, such as signal generation, and measurement of stator voltages and data outputs. The position is available as an absolute position or as a simulation of incremental rotary encoders.

The resolver signals are shown in figures 7.21, 7.22 and 7.23.

The resolver is an absolute sensor, but only within one revolution. If the position within more revolutions has to be measured, a counter has to be added. The sensor then becomes a relative sensor.

A typical accuracy of the resolver is around 0,05% of range (360°), the typical resolution is 4096 positions per one revolution.

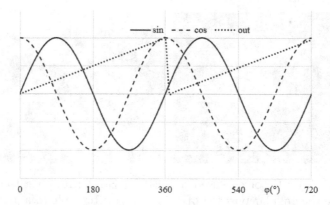

FIGURE 7.23: Amplitudes of SIN and COS signals in dependence on position, the resolver-to-digital output is linear. The resolver is an absolute sensor in range of one revolution

7.3 Capacitive sensors

Capacitive sensors are based on capacity changes of the sensing element. The change of capacity can be caused by the change of three variables. The change of capacity can be caused by the change of 3 variables, distance between the electrodes d, electrode area A and/or the change of permittivity ϵ_0. All three options are used for sensors.

Relative permittivity for selected materials is shown in table 7.1.

For a planar capacitor, the capacity is

$$C = \varepsilon_0 \varepsilon_r \frac{A}{d} \tag{7.20}$$

where A is the electrode area, d is the electrode distance, ϵ_0 is vacuum permittivity ($8.854187817... \times 10^{-12}$ A $\cdot$ s/(V $\cdot$ m) $= 8.854187817... \times 10^{-12}$ F/m), and ϵ_r is relative permittivity (material property)

TABLE 7.1: Relative permittivity of selected materials [39].

Material	Relative permittivity [-]
vacuum	1 (by definition)
air	1.00058986 ± 0.00000050
paper	3.5
graphite	10–15
silicon	11.68
water	88, 80.1, 55.3, 34.5(0, 20, 100, 200°C)

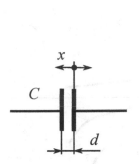

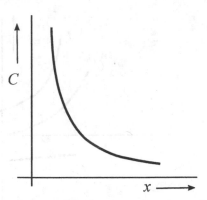

FIGURE 7.24: Capacitive sensor with change of distance between electrodes [6]

FIGURE 7.25: Steady-state characteristics of a capacitive sensor with change of distance between electrodes [6]

a) Sensors based on change of distance between electrodes
Sensitivity [6]

$$C = \varepsilon_0 \varepsilon_r \frac{A}{d(x)} \qquad \frac{\Delta C}{\Delta d} \cong \frac{-C}{d}\left[1 - \frac{\Delta d}{d}\right] \qquad (7.21)$$

One electrode is fixed, attached to a frame—figure 7.24. The other is attached to the measured object whose position should be measured. It can also be the object itself. As the electrode is moving, the capacity changes are proportional to the changes in position. However the dependence is non-linear, as shown in figure 7.25.

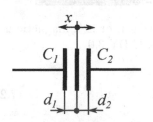

FIGURE 7.26: Differential capacitor [6]

b) Differential capacitor
The simple sensor with change of distance between the electrodes has a nonlinear steady-state characteristic. This can be inconvenient. In order to obtain better linearity and higher sensitivity a differential capacitor is often used.

The change of capacity is not evaluated between two electrodes but between two fixed and one movable electrode. The movable electrode is placed between the fixed ones, as shown in figure 7.26. The circuit is connected as a bridge connection. The resulting connection yields a better linearity as shown in figure 7.27.

An example of a differential capacitive sensor in a magnetic field sensor is shown in figure 7.28.

$$C_1 = \varepsilon_0 \varepsilon_r \frac{A}{d_1(x)} \qquad (7.22)$$

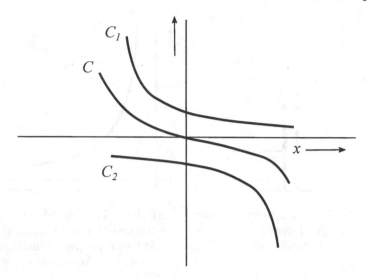

FIGURE 7.27: Steady-state characteristic of a differential capacitive position sensor [6]

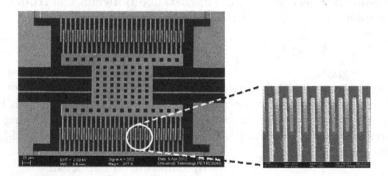

FIGURE 7.28: Example of a differential capacitive sensor in a magnetic field sensor, courtesy of [40], license CC BY 3.0

$$C_2 = \varepsilon_0 \varepsilon_r \frac{A}{d_2\,(x)} \qquad (7.23)$$

Sensitivity [6]

$$\frac{\Delta C}{\Delta d} \cong \frac{-C}{d}\left[1 - 2\left(\frac{\Delta d}{d}\right)^2\right] \qquad (7.24)$$

c) **Change of electrode area** In this arrangement, one electrode is fixed to a frame, as shown in figure 7.29. The other electrode is movable, attached to the object or it is the object itself. By movements of the electrode, the area is changing and hence the capacity is a function of position. The change is

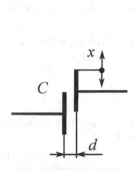

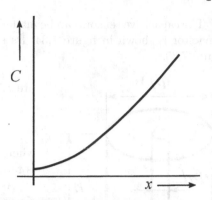

FIGURE 7.29: Capacitive sensor
with change of area [6]

FIGURE 7.30: Steady-state
characteristic of a capacitive
sensor with area change [6]

linear as shown in figure 7.30.

$$C = \varepsilon_0 \varepsilon_r \frac{A(x)}{d} \qquad (7.25)$$

Sensitivity [6]

$$\frac{\Delta C}{\Delta l} \cong \frac{-C}{d} \left[1 + \frac{\Delta d}{d} \right] \qquad (7.26)$$

d) Change of permittivity

This principle is based on the change of permittivity between the electrodes, as shown in figure 7.31. Most often the material is changed, i.e., air is replaced with a liquid in a liquid-level sensor. The static characteristic is linear as shown in figure 7.32.

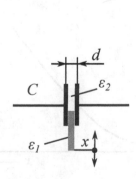

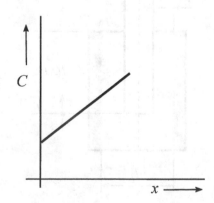

FIGURE 7.31: Capacitive sensor
with permittivity change [6]

FIGURE 7.32: Steady-state
characteristic of a capacitive
sensor with permittivity change [6]

The capacitive sensors can be made as planar or cylindrical. The cylindrical capacitor is shown in figure 7.34. Its steady-state characteristic is shown in figure 7.35.

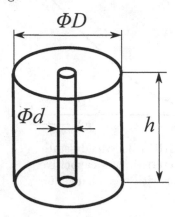

ΦD

Φd

h

FIGURE 7.33: Cylindrical capacitor

For a cylindrical capacitor, shown in figure 7.33, the capacity is

$$E = \frac{\lambda}{2\pi \cdot \varepsilon \cdot r} \qquad (7.27)$$

where λ is the charge per unit length and r is radius ($r = D/2$).

Assuming the diameter of the central electrode d is significantly smaller than the diameter of the external electrode D then

$$\Delta V = \frac{\lambda}{2\pi \cdot \varepsilon} \int_d^D \frac{1}{r} dr = \frac{\lambda}{2\pi \cdot \varepsilon} \ln\left(\frac{D}{d}\right) \qquad (7.28)$$

$$C = \frac{2\pi \cdot \varepsilon}{\ln\left(\frac{D}{d}\right)} \cdot h \qquad (7.29)$$

Capacity

$$C = \varepsilon \frac{2\pi l\left(x\right)}{\ln\left(\frac{d_1}{d_2}\right)} \qquad (7.30)$$

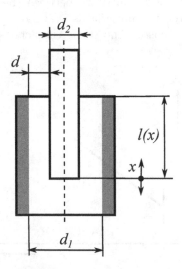

d_2

d

$l(x)$

x

d_1

FIGURE 7.34: Cylindrical capacitive sensor[6]

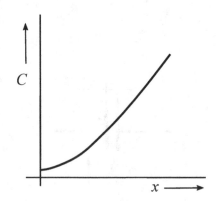

C

$x \longrightarrow$

FIGURE 7.35: Steady-state characteristic of a cylindrical capacitive sensor [6]

Sensitivity [6]

$$\frac{\Delta C}{\Delta l\left(x\right)}\tilde{=}\frac{C}{d}\left[1-\frac{1}{2}\left(\frac{\Delta d}{d}\right)^{2}\right] \tag{7.31}$$

Capacitive sensors are among the most precise position sensors. Very small distances can be measured, in the order of nm. For comparison, the size of an atom is around 0,1 to 0,5 nm (function of the element).

7.4   Magnetic (Hall) sensors

Hall sensors are based on the Hall principle. They allow sensing the presence of a magnetic field in a semiconductor. They are typically used as end switches on production lines, speed sensors in automotive, etc.

The passage of a magnetic field is sensed with a Hall sensor. The position sensor principle is shown in figure 7.36. The magnetic field is often created with magnetized areas on a magnetic strip or tape. The tape is fixed to the machine frame, and the Hall sensor moves above it. The spacing of the magnetic marks on the tape is known, hence the position can be calculated. The magnetic sensors can be absolute or relative. Magnetic sensors can also be used as speed sensors. In this case, the frequency of the pulses is counted. Magnetic sensors are very robust and reliable. Examples of magnetic sensors are shown

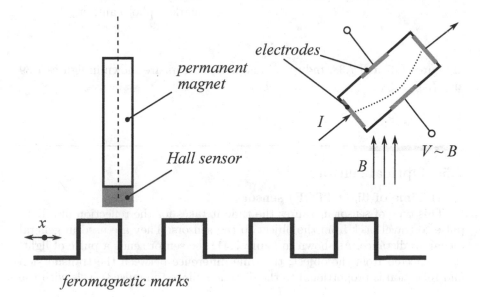

FIGURE 7.36: Principle of a Hall sensor

FIGURE 7.37: Magnetic field   FIGURE 7.38: Magneto-resistive
sensor TLE4905G—photo author   sensor KMZ10C—photo author

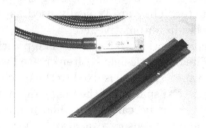

FIGURE 7.39: Magnetic position   FIGURE 7.40: Linear magnetic
sensor (top), tape with magnetic   displacement sensor, black strip is
marks (bottom)—photo author   the tape with magnetic
                                         marks—photo author

in figures 7.37 and 7.38. Industrial magnetic sensors are shown in figures 7.39 and 7.40.

7.5 Optical sensors

a) Time of flight (TOF) sensors

This type of sensor measures the time it takes for the reflection of a light pulse to travel back from the object to the sensors. They are used in general for larger distance. As shown in figure 7.41, the sensor sends a pulse of light. It is reflected from the object; the time difference between the transmission and reflection is proportional to the distance. Since the pulse travels with the

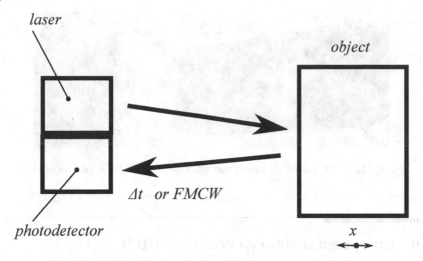

FIGURE 7.41: Optical distance sensor—TOF

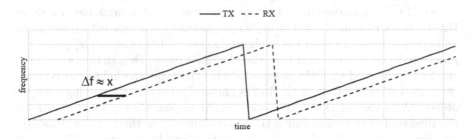

FIGURE 7.42: FMCW principle

speed of light, the time difference is very short and difficult to measure for small distances.

b) Frequency modulated continuous wave (FMCW)

This principle eliminates the measurement of small time differences in TOF sensors. The transmitter, in figure 7.42, sends a continuous wave, a signal with variable frequency. The frequency of the transmitted wave is compared with the frequency of the received wave. The received wave is reflected from the measured object. The frequency difference is proportional to the measured distance. Frequency difference can be measured easily with a demodulator. The maximal measured distance is limited by the frequency sweep available. An example where FMCW is used is a laser distance meter shown in figure 7.43.

FIGURE 7.43: Laser distance meter, use of FMCW—photo author

7.6   Incremental rotary encoders (IRC)

Incremental rotary encoders are among the most accurate and often-used sensors. They are used, for example, in machine tools or measurement devices.

An IRC is an optical digital sensor. The construction principle of a linear IRC is shown in figure 7.44. The main components are a tape with transparent and opaque fields and two LED + photo-diode (or photo-transistor) pairs. The photo-diode detects the light from the LED as it passes though the tape. When the LED is aligned with the opaque field on the tape, the light does not pass through. Either the tape or the LED+photo-diode pairs are moving with the object. With one LED+photo-diode pair, it is possible to detect speed but not direction. The second pair is therefore shifted from the first one. From the two output signals, both speed and direction can be measured. The output signals are typically called A and B. The AB output signal is show in figure 7.45. Signals A and B are typically shifted by a quarter-period. The measured

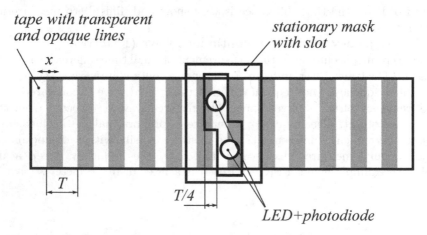

FIGURE 7.44: Principle of linear incremental encoder

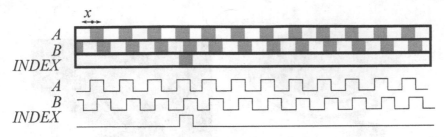

FIGURE 7.45: Linear IRC scale and output signals

distance is calculated from the known resolution of the scale and the number of pulses in an external counter circuit.

The achieved resolution is typically in the range of 5 to 10 µm although more or less accurate sensors are available to suit the application needs. The linear IRC suffers from thermal expansion of the scale. The rotary version, shown in figure 7.46, does not suffer from this. A rotary IRC is composed from a glass or plastic disc as opposed to the linear IRC with a tape. Other components such as the LED+photo-diode pairs are similar and have the same function. A typical, easily achievable resolution is 10000 pulses/revolution or more.

An IRC is a relative sensor. After power-up, the absolute position is not available. Another device is required to set the reference position, such as an end switch. Devices equipped with relative sensors require homing after power up. On the other hand, relative sensors have typically higher resolution than absolute rotary encoders.

Figure 7.47 shows an IRC used to measure the angular position of a tram wheel. Another industrial example is an IRC in a torque sensor shown in

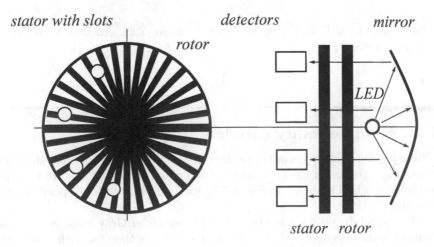

FIGURE 7.46: Incremental rotary encoder

FIGURE 7.47: An IRC used to measure the angular position of a tram wheel—photo author

FIGURE 7.48: An IRC in a torque sensor—photo author

FIGURE 7.49: IRC code disc, photo author

FIGURE 7.50: IRC—photo author

figure 7.48. An IRC code disc is shown in figure 7.49. An industrial IRC is shown in figure 7.50.

7.7 Absolute rotary encoders

An absolute rotary encoder operates in a similar way as an IRC. However, it has more tracks on the code wheel. Each track encodes 1 bit. An example of the code wheel is shown in figure 7.51 for an 2-bit encoder (4 positions per revolution).

The absolute position is encoded in Gray code. Gray code is a special form of binary encoding; it changes only 1 bit at a time for each transition.

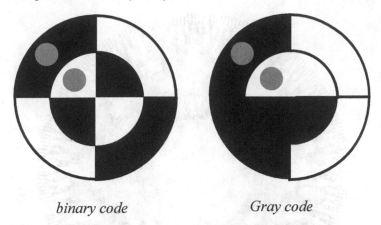

binary code   Gray code

FIGURE 7.51: Code disc for a 4=bit absolute rotary encoder—left with binary code, right with Gray code

TABLE 7.2: Binary vs. Gray code for 2 bits

Binary	Gray
00	00
01	01
10	11
11	10

This allows simple error detection. An example of Gray code for 2 bits is shown in table 7.2.

The available space in the encoder housing limits the achievable resolution. Therefore absolute rotary encoders have typically smaller resolution and are more expensive than relative IRCs.

Absolute rotary encoders have typically a resolution of 2^{10} to 2^{12} bits = 1024 to 4096 positions / revolution. An example of a code wheel for an absolute position encoder with resolution of 10 bits (1024 positions/revolution), Gray code, is shown in figure 7.52.

7.8 Microwave position sensor (radar)

This sensor type is used to measure both large and short distances. Based on the required range the suitable principle is selected. For short distances, Frequency Modulated Continuous Wave (FMCW); for large distances, Time of Flight (TOF).

FIGURE 7.52: Code wheel for an absolute rotary encoder, 10 bit (1024
positions/revolution), Gray code, generated with [41]

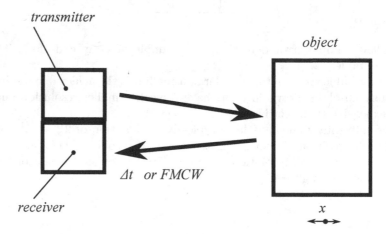

FIGURE 7.53: Microwave position sensor

In both cases, radio wave reflection from the measured object is detected,
as shown in figure 7.53.

a) Time of Flight (TOF) method

This method is used to measure larger distances. The transmitter sends a
microwave pulse. The time it takes for the reflection from the object to get

back to the sensor is timed. Since the signal propagates with the speed of light, the time difference is very small (order of fs), and very accurate measurement of time is required.

b) Frequency-Modulated Continuous Wave (FMCW) method

The measurement of very small time difference in the TOF method is not necessary for FMCW. The frequency difference is measured instead. The principle is shown in figure 7.54. The transmitter sends out a continuous wave, and its frequency is changed. The signal gets reflected from the object. The frequency of the received signal is compared to the frequency of the transmitted signal. The frequency difference is proportional to the measured distance. The further the object is, the longer it takes for the signal to travel and hence the larger the frequency difference. The frequency difference is obtained from a mixer circuit as shown in the block diagram shown in figure 7.55.

The maximal range is limited by the maximal allowed frequency sweep. The used frequency ranges are 9 to 10 GHz, 24 GHz and 26GHz. FMCW position, distance or liquid-level sensors can be used for both conductive and dielectric material. The achievable resolution is around 1 mm. Examples of radar sensors are shown in figures 7.56 and 7.57.

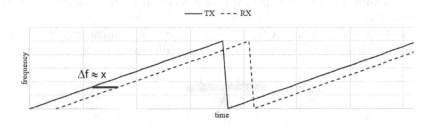

FIGURE 7.54: Principle of FMCW

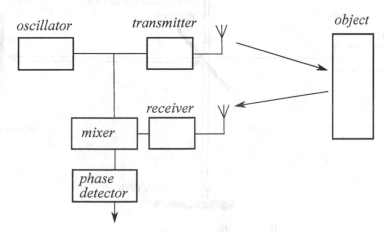

FIGURE 7.55: FMCW block diagram

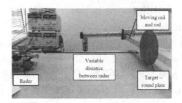

FIGURE 7.56: Transmitter and   FIGURE 7.57: Radar sensor,
receiver for a radar sensor,   courtesy of [42], license CC BY 4.0
courtesy of [42], license CC BY 4.0

7.9   Interferometers

An interferometer is an optical instrument. It uses constructive and destructive interference to achieve a very fine resolution. The achieved resolution is up to 0,1 μm. The principle of Michelson's interferometer is shown in figure 7.58.

The light source sends a light beam towards a semitransparent mirror. The beam splits in to two beams. One beam is directed towards the measured

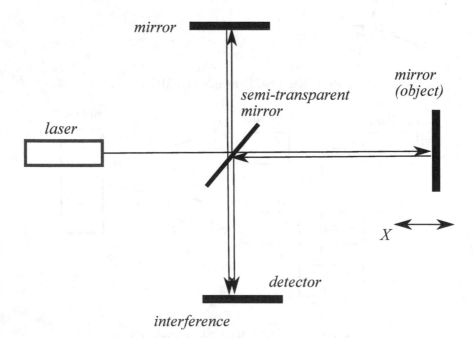

FIGURE 7.58: Principle of Michelson's interferometer

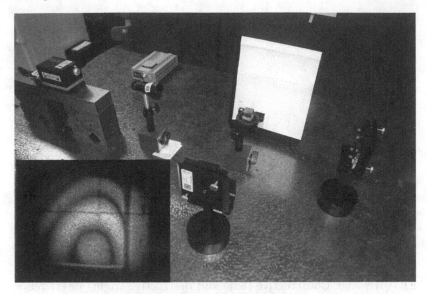

FIGURE 7.59: Michelson's interferometer, bottom left interference fringes on the detector—photo author

object, the second towards a fixed mirror. The measured object is equipped with a mirror as well. The object moves in the x direction in figure 7.58. The beam from the object and the fixed mirror reflects back. Both beams interfere on the detector. An experiment with the interferometer is shown in figure 7.59. An example of interference is shown in figure 7.60. As the object moves, the interference is either constructive or destructive, and dark or bright strips are visible on the detector. By counting the number of transitions, the object's position can be very accurately measured.

7.10 Proximity sensors

Proximity sensors detect only the presence of an object. They do not detect its distance. The output of a proximity sensor is a logic signal, 1 or 0. Proximity sensors are significantly less expensive than continuous position sensors. Proximity sensors are used, for example, on production and packing lines to detect objects. They can also be used as speed sensors.

7.10.1 Inductive proximity sensors

Inductive proximity sensors use a coil whose inductance is changed by the detected object. The coil is powered with AC current. When a object enters

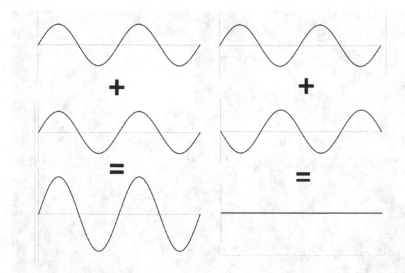

FIGURE 7.60: Constructive (left) and destructive (right) interference

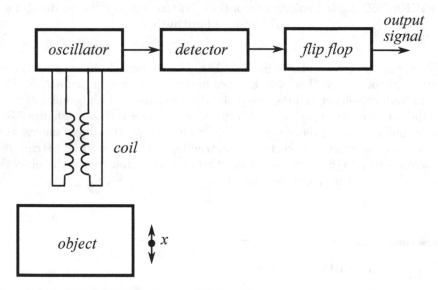

FIGURE 7.61: Principle of inductive proximity sensor

the detection zone, the inductance changes and is detected. The principle is shown in figure 7.61. The output signal is 1 or 0.

Inductive proximity sensors work only with conductive objects and preferably with ferromagnetic objects. They cannot detect non-conductive objects, such as plastic, glass, wood etc.

TABLE 7.3: Sensitivity of an inductive sensor for selected material. Sn = detection distance, based on [43]

Material	Sensing distance
Fe37 (Iron)	$1 \times$ Sn
Steel	$0.9 \times$ Sn
Brass, bronze	$0.5 \times$ Sn
Al	$0.4 \times$ Sn
Cu	$0.4 \times$ Sn

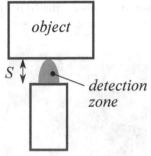

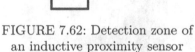

FIGURE 7.62: Detection zone of an inductive proximity sensor

FIGURE 7.63: Inductive proximity sensor—photo author

The detection distance is a function of sensor diameter. The larger the diameter the larger the detection distance. The detection distance is in the order of mm, up to approximately 1 cm.

The detection distance is also a function of the material of the detected object. Best results (largest detection distance) are achieved for iron (ferromagnetic). The detection distance decreases for other conductive materials, as shown in table 7.3.

Examples of industrial inductive proximity sensors and their applications are shown in figures 7.63, 7.64, 7.65, 7.66, 7.67, 7.68, 7.69.

7.10.2 Capacitive proximity sensors

The principle of a capacitive proximity sensor is shown in figure 7.70. The inter-electrode capacity is changed as the object approaches the sensor. The capacity change alters the frequency of the oscillator; this gets detected by the detector. The flip-flip outputs the output signal, a 1 or 0.

The capacitive sensor detects conductive and non-conductive objects. The industry traditionally uses inductive sensors for metallic objects and capacitive sensors for non-metallic objects. Capacitive sensors, however, detect both.

FIGURE 7.64: Inductive
proximity sensor—photo author

FIGURE 7.65: Inductive
proximity sensor—photo author

FIGURE 7.66: Inductive
proximity sensor on a barrel filling
line—photo author

FIGURE 7.67: Inductive
proximity sensor—photo author

FIGURE 7.68: Inductive
proximity sensor—photo author

FIGURE 7.69: Inductive
proximity sensor used as a speed
sensor—photo author

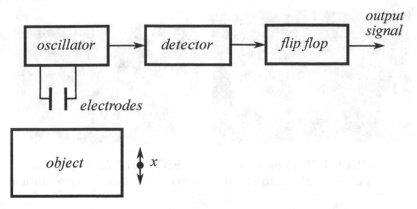

FIGURE 7.70: Principle of capacitive proximity sensor

TABLE 7.4: Sensitivity of capacitive proximity sensors for selected materials. Sn = detection distance - based on [44]

Material	Sensing distance
Metals	$\sim 1 \times$ Sn
Water	$\sim 1 \times$ Sn
Plastics	$\sim 0.5 \times$ Sn
Glass	$\sim 0.5 \times$ Sn
Wood	$\sim 0.4 \times$ Sn

The detection distance is in the order of mm, up to approximately 1 cm. It is influenced by the objects permittivity. Best detection distance is for conductive materials such as metals, water, etc. The detection distance decreases for other materials as shown in table 7.4.

The detection zone has the same shape as for an inductive proximity sensor shown in figure 7.62.

Examples of industrial capacitive proximity sensors and their applications are shown in figures 7.71, 7.72, 7.73, 7.74.

7.10.3 Optical proximity sensors

Optical proximity sensors are based either on the reflection of a light beam from the object, figure 7.75, or on the interruption of a light beam by the object. Optical sensors are typically not suitable for transparent objects.

An optical proximity sensor (continuous) is shown in figure 7.76. A PIN photodiode used as a detector is shown in figure 7.77. An optical proximity sensor is shown in figures 7.80 and 7.81. An optical proximity sensor used on an automated assembly line is shown in figure 7.82, mounted on a conveyor belt it is shown in figure 7.83. A reflective sensor is shown in figure 7.78. An optical proximity sensor with an optical fiber is shown in figure 7.79.

FIGURE 7.71: Capacitive
proximity sensor—photo author

FIGURE 7.72: Capacitive
proximity sensors—photo author

FIGURE 7.73: Proximity sensors,
from left: Hall, inductive,
capacitive, optical—photo author

FIGURE 7.74: Capacitive
proximity sensor—photo author

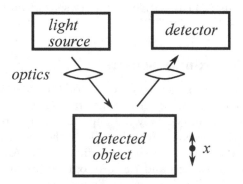

FIGURE 7.75: Optical proximity sensor—reflection from the detected object

FIGURE 7.76: Optical proximity
sensor (continuous)
HSDL-9100—photo author

FIGURE 7.77: PIN photodiode
SFH203—detector for optical
sensors—photo author

FIGURE 7.78: Reflective sensor
TCRT1000—photo author

FIGURE 7.79: Optical proximity
sensor with an optical
fiber—photo author

FIGURE 7.80: Optical proximity
sensor—photo author

FIGURE 7.81: Optical proximity
sensor—photo author

FIGURE 7.82: Optical proximity
sensor—photo author

FIGURE 7.83: Optical proximity
sensor on a conveyor belt—photo
author

8

Speed and RPM

1. Mechanical—use directly a mechanical principle such as friction. The value is shown on a scale. One example is a mechanical speedometer in older cars. Mechanical speed sensors may have an electric output signal useful for further processing.

2. Electromagnetic—i.e., tachodynamo. It is in principle an electric generator modified for measurement, with an electric signal output.

3. Optoelectronic/magnetic—in principle a proximity sensor that detects the passage of marks on a moving object.

4. Capacitive/inductive—same as previous category, they detect the passage of marks.

8.1 Electromagnetic—tachodynamo

Tachodynamo is an electric machine. Its construction is shown in figure 8.1. It is a DC electric generator. The output voltage is proportional to speed.

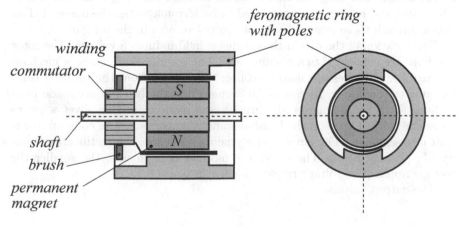

FIGURE 8.1: Tachodynamo

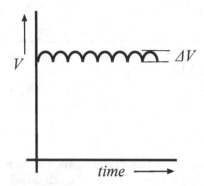

FIGURE 8.2: Voltage ripple on tachodynamo output

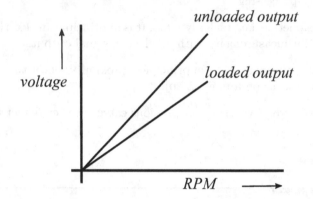

FIGURE 8.3: Static characteristic of a loaded and unloaded tachodynamo

The source of magnetic flux (permanent magnet or field winding) is placed on the stator. The magnetic flux is closed through the air gap and the rotor. The stator and rotor components need to be ferromagnetic, the magnetic flux closes through those materials. A coil (rotor) rotates in the air gap.

The rotation of the coil in the magnetic field induces a voltage. The rotor windings are connected to a commutator. The commutator acts as a mechanical rectifier. Brushes are used to collect the energy from the rotating commutator to the stator. In case of DC generators, the brushes are made from carbon. For a tachodynamo, the brushes are made from thin silver wires to lower the voltage drop on the brush-commutator contact. The commutator is split into segments. The number of segments influences the rectification quality, the voltage ripple. The larger the number of segments, the smaller the voltage ripple. The voltage ripple is shown in figure 8.2.

The output voltage is

$$V_o = C_{SS} \cdot \Phi \cdot \omega \tag{8.1}$$

where C_{SS} is a constant describing the machine construction, Φ is magnetic flux and ω is angular speed.

Tachodynamo has a voltage output. The ideal load resistance is therefore $R = \infty$. If some current is taken from the output, a voltage drop is created on the internal resistance. This has an effect on the static characteristic, as shown in figure 8.3. Therefore, the minimal allowed load resistance is specified by manufacturers.

The limited number of commutator sections leads to an imperfect rectification. The voltage ripple is typically around 2% of the measured voltage. This limits the achievable resolution. If speed resolution better than 1% or 2% is required, the tachodynamo is not a suitable sensor. Other sensors such as IRC or an inductive speed sensor should be used in this case.

A tachodynamo used as a motor speed sensor is shown in figure 8.4. The cut-away view of a tachodynamo is shown in figure 8.5. A tachodynamo used as speed sensor on a dynamometer is shown in figure 8.6. Tachodynamo as a speed sensor in optical fiber production is shown in figure 8.7.

FIGURE 8.4:
Tachodynamo—photo author

FIGURE 8.5: Tachodynamo,
cutaway view—photo author

FIGURE 8.6: Tachodynamo
(bottom right) and inductive
sensor (center)—speed sensing on
dynamometer—photo author

FIGURE 8.7:
Tachodynamo—speed sensing
during production of optical
fibers—photo author

8.2   Optoelectronic speed sensors

An optoelectronic speed sensor is shown in figure 8.8. The sensor is composed of a disk with holes or teeth, creating "marks." On one side of the disk there is a light source (LED, bulb); a detector (photodiode, photo-transistor) is placed on the other side. The detector detects the passage of light through the mark (hole, teeth gap).

An example of this principle is shown in figure 8.9, where an old computer mouse is shown. The two black disks are the disks with gaps and "marks." A similar example, a fan speed sensor, is shown in figure 8.10.

The output signal is a rectangular voltage signal. Its frequency is proportional to the measured speed. Using the known number of marks on the disc, the speed can be calculated. When, for example, 60 marks are used, the frequency displayed on a frequency counter is equal to the speed in RPM.

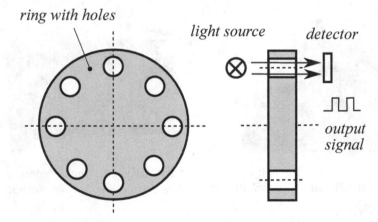

FIGURE 8.8: Principle of optoelectronic speed sensor

FIGURE 8.9: Optical sensor, old computer mouse—photo author

FIGURE 8.10: Optical speed sensor—photo author

8.3 Inductive speed sensors

The sensor, shown in figure 8.11, is composed of a source of magnetic flux (permanent magnet) and a sensing coil. When a ferromagnetic object enters the magnetic field, the magnetic flux changes. Changes in magnetic flux cause induced voltage in the sensing coil. The change of magnetic flux is achieved by ferromagnetic "marks" on the sensed object. The marks are typically created as teeth and gaps.

The frequency (and amplitude) of the induced voltage is proportional to speed. Since the amplitude is also a function of the air gap between the sensor and the object and frequency is not, frequency detection is used in speed sensing. As the induced voltage is a function of speed, it is small for small speeds. Therefore, the inductive speed sensor can't work from zero speed. It has always some minimal speed.

The FEM calculation of magnetic flux around a inductive speed sensor is shown in figure 8.12. The changes of magnetic flux create induced voltage, as shown in figure 8.13. Examples of industrial inductive speed sensors are shown in figures 8.14 and 8.15.

An inductive speed sensor can be used for angular speed or for linear speed as well. In case of linear speed, the source of "marks" is a linear rod.

Inductive speed sensors are used, for example, in automotive as ABS speed sensors to measure the speed of wheels.

Inductive sensors are very sensitive to the correct distance to the object. The sensing distance is in the order of few mm.

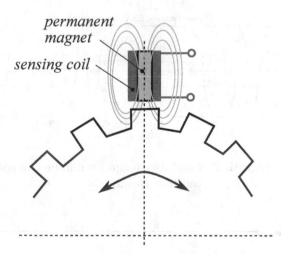

FIGURE 8.11: Principle of inductive speed sensor

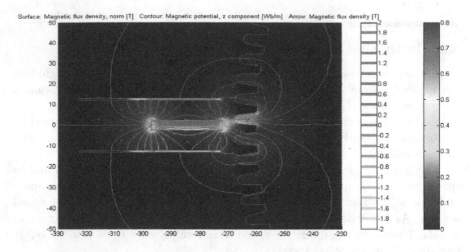

FIGURE 8.12: Magnetic flux around an inductive speed sensor, FEM calculation, [45]

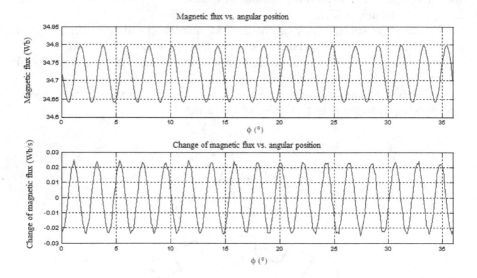

FIGURE 8.13: Magnetic flux and its change for a inductive speed sensor for constant speed, [45]

8.4   Hall speed sensors

The Hall speed sensor is used in a similar way as the inductive sensor. It measures changes in the magnetic flux. The magnetic flux can come from a

FIGURE 8.14: Inductive speed
sensor (bottom left), Hall speed
sensor (center right)—photo
author

FIGURE 8.15: Two inductive
speed sensors—photo author

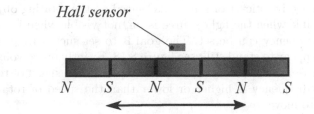

FIGURE 8.16: Hall speed sensor, linear—source of magnetic flux is a
magnetized tape with N and S poles

permanent magnet (figure 8.17), from a magnetic tape (figure 8.16) or from
the measured object itself.

Hall speed sensor is a robust speed sensor. It is used in similar applications as the inductive speed sensor such as an ABS wheel speed sensor. It is,
however, capable of detecting speed from zero.

The detection distance is in the order of a few mm; the sensor is sensitive
to this distance. When the distance is not set properly, the sensor starts to
drop the output pulses. This leads to false speed readings. The maximal pulse
frequency is typically in the range of 10 to 15 kHz.

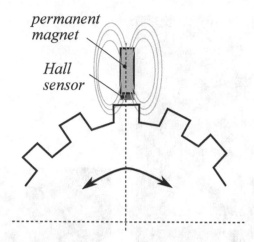

FIGURE 8.17: Hall speed sensor, angular—source of magnetic flux is a permanent magnet build into the sensor; the object is ferromagnetic

8.5   Stroboscopes

A stroboscope is a handheld instrument used to measure RPM. Its principle is shown in figure 8.18. It uses a source of light flashes with variable frequency. A reflective sticker or a mark is placed on the rotating object. The reflection is visible when the light source is on, not visible when it's off.

The flash frequency can be set. The goal is to set such a frequency that the mark will appear to stand still; see figure 8.18. Exactly one revolution was made. In this case the frequency of flashes is equal to the speed of rotation.

If the flash frequency is higher or lower than the speed of rotation, the mark appears to move.

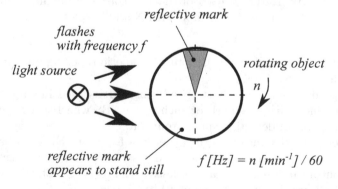

$$f\,[Hz] = n\,[min^{-1}]\,/\,60$$

FIGURE 8.18: Stroboscope—the mark appears to stand still

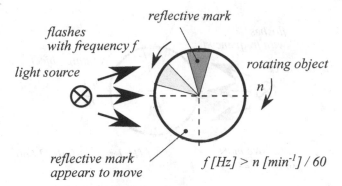

FIGURE 8.19: Stroboscope—the mark appears to move against the direction of rotation

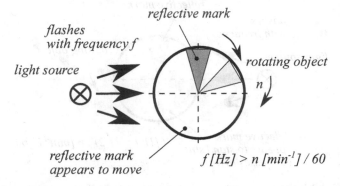

FIGURE 8.20: Stroboscope—the mark appears to move in the direction of rotation

If the speed of rotation is smaller than the flash frequency, the mark appears to rotate against the direction of rotation. This is shown in figure 8.19.

If the speed of rotation is larger than the flash frequency, the mark appears to rotate in the direction of rotation. This is shown in figure 8.20.

When the flash frequency is exactly double the speed of rotation, two marks are visible. The marks appear to stand still. This is shown in figure 8.21.

When the flash frequency is exactly 3× the speed of rotation, three marks are visible. The marks appear to stand still. This is shown in figure 8.23.

Both situations can be easily recognized, since it is known only one mark exists.

However, when the flash frequency is exactly half the speed of rotation, one mark is visible. The mark appears to stand still, but it is visible only every second revolution. This is shown in figure 8.22. It's impossible to recognize this situation.

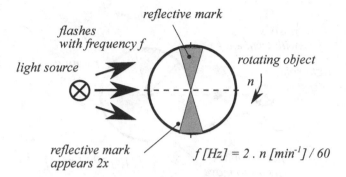

FIGURE 8.21: Stroboscope—two marks are visible, flash frequency is 2× of speed of rotation

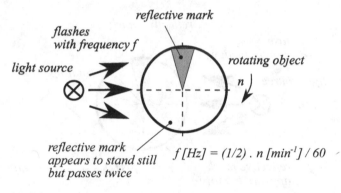

FIGURE 8.22: Stroboscope—flash frequency is half the speed of rotation, the mark passes 2× before it is visible; if only one is visible = wrong setting

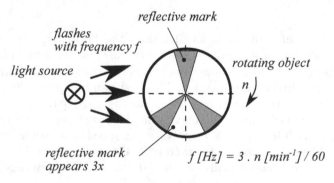

FIGURE 8.23: Stroboscope—3× flash frequency, three marks are visible

Therefore, the stroboscope user should always start from the maximal flash frequency to avoid this problem.

9

Acceleration

Acceleration sensors measure the effect of inertia on a mass, as evident from Newton's second law

Force = - mass . acceleration

$$F = -m \cdot a \tag{9.1}$$

The effects of acceleration can be measured in many ways, for example, measure the force directly with strain gauges, or measure the position change with a position sensor. Those principles work in a different frequency range, as shown in figure 9.1.

9.1 Piezoelectric acceleration sensors

The piezoelectric acceleration sensor measures the piezoelectric voltage generated from a piezo crystal that is being pushed by the inertial force. Main sensor parts are shown in figure 9.2. A seismic mass is connected to the piezo crystal. The generated voltage is proportional to the measured acceleration.

Electrodes are deposited on the piezo crystal. The voltage is produced only when the crystal is put into compression or tension, not statically. The voltage is not produced with a static force. Therefore, this type of sensor does not measure from low frequencies. It is suitable to measure vibrations

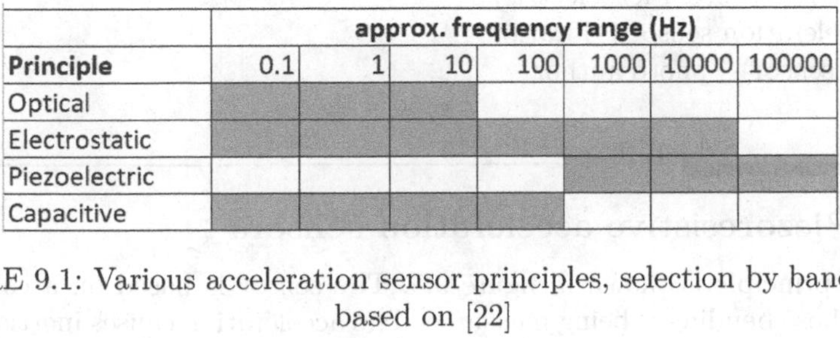

Principle	approx. frequency range (Hz)						
	0.1	1	10	100	1000	10000	100000
Optical							
Electrostatic							
Piezoelectric							
Capacitive							

FIGURE 9.1: Various acceleration sensor principles, selection by bandwidth; based on [22]

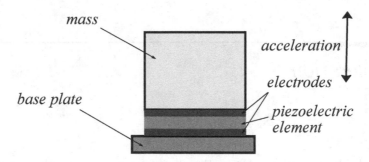

FIGURE 9.2: Principle of piezoelectric acceleration sensor

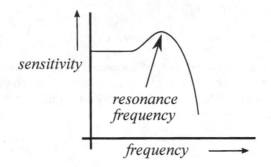

FIGURE 9.3: Typical frequency response of a piezoelectric acceleration
sensor, response with resonance

FIGURE 9.4: Piezoelectric
acceleration sensor (vibration
sensor)—photo author

but not a constant acceleration. The
electronic circuit requires a high-
input resistance amplifier (few GΩ).

The maximal measured frequency
is limited by the mechanical resonance, as shown in figure 9.3. The
piezoelectric acceleration sensor is
shown in figure 9.4.

9.2   Piezoresistive acceleration sensors

The principle is shown in figure 9.5. The sensor is based on a cantilever
beam whose bending is being measured. The acceleration causes inertial forces
on the central mass. The mass moves up or down, based on the direction of
the acceleration. The flexing (position) of the cantilever beam is measured.

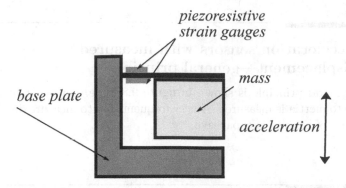

FIGURE 9.5: Piezoresistive acceleration sensor

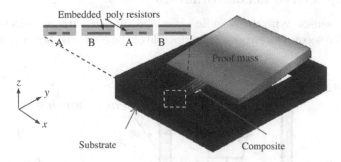

FIGURE 9.6: Piezoresistive MEMS accelerometer—courtesy of [46], license CC BY 3.0

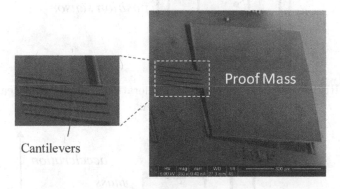

FIGURE 9.7: Piezoresistive MEMS accelerometer—courtesy of [46], license CC BY 3.0

The flexing is measured with a piezoresistive force sensor on the cantilever. Its deformation is proportional to the measured acceleration. The structure is shown in figure 9.6, a microscope photograph of a MEMS accelerometer is shown in figure 9.7.

9.3   Acceleration sensors with measured displacement—general principle

The general principle is shown in figure 9.8. The movement of a seismic mass due to inertia is measured. A very frequent way to measure displacement in this kind of sensors is capacitive.

9.4   Capacitive accelerometers

In this sensor type the seismic mass is equipped with one electrode, the frame with a second electrode. The principle of the sensor is shown in figure 9.9.

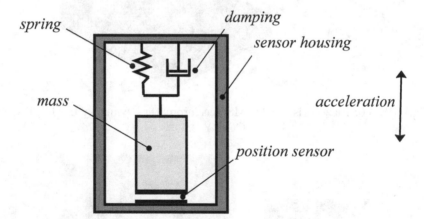

FIGURE 9.8: General principle of acceleration sensors with measured displacement

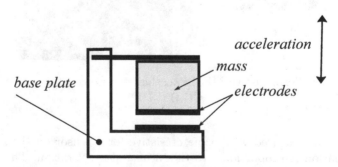

FIGURE 9.9: Capacitive accelerometer

The electrodes form a differential capacitor. When the seismic mass moves with the applied acceleration, the capacity changes.

Due to better linearity, not a single capacitor's capacity is measured, but the measurement is differential. The capacity is measured between the movable and fixed parts as a differential capacitor, shown in figure 9.10. The capacitive accelerometer, produced as a Micro Electro Mechanical System (MEMS) is a very common sensor today. It is used in many applications, such as airbag control, inertial measurement units, etc. Capacitive accelerometer chips are shown in figure 9.11, a detail of the differential capacitor under a microscope is shown in figure 9.12. The internal structure and microscope photograph of such an accelerometer is shown in figure 9.13.

FIGURE 9.10: Differential capacitor

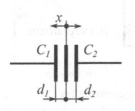

FIGURE 9.11: Capacitive accelerometers—photo author

FIGURE 9.12: Capacitive accelerometer—courtesy of [47], license CC BY 4.0

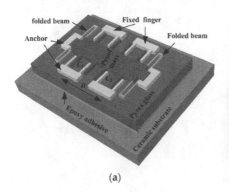

(a)

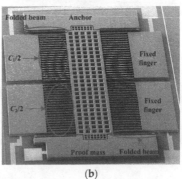

(b)

FIGURE 9.13: Capacitive accelerometer—courtesy of [48], license CC BY 4.0

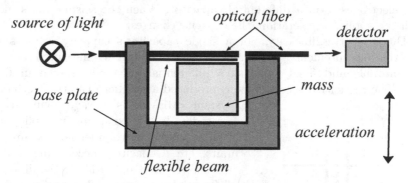

FIGURE 9.14: Principle of optical accelerometer

9.5   Optical accelerometers

The principle of an optical accelerometer is shown in figure 9.14. The optical fiber is supplied with a light source. The optical fiber ends on a seismic mass on a cantilever beam. The optical fiber continues as the output optical fiber after the seismic mass. When the acceleration is applied, the seismic mass moves. This changes the coupling between the input and output optical fibers. A larger movement further attenuates the signal. The mutual movements are in the order of hundreds of μm [49].

This accelerometer is suitable for explosive environments as no electrical signals are being used. It is also suitable for environments with strong magnetic fields.

10

Pressure

The definition of pressure can be made using two equivalent definitions.

a) Force acting on a given area (in normal direction to the surface)—see figure 10.1, left.

Pressure is then

$$p = \frac{F}{A} = \frac{m \cdot g}{A} \quad [Pa; kg; m \cdot s^{-2}; m^2] \tag{10.1}$$

b) Hydrostatic pressure See figure 10.1 right

$$p = \rho \cdot g \cdot h \quad [Pa; kg \cdot m^3; m \cdot s^{-2}; m] \tag{10.2}$$

The pressure can be measured in several ways:

Absolute pressure—p_a—measured from zero pressure (vacuum)

Relative pressure—measured from a selected reference. The reference can be, for example, barometric pressure p_b. Over- or under-pressure is measured. Differential pressure gauges are used.

In flowing fluids, the pressure also depends on the flow velocity. Total pressure = static pressure (for flow velocity 0) + kinetic pressure.

Kinetic pressure is

$$p_d = \frac{\rho \cdot w^2}{2} \tag{10.3}$$

where w is flow velocity.

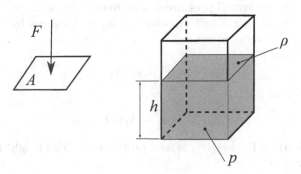

FIGURE 10.1: Pressure definition: (left) force acting on area, in the normal direction; (right) hydrostatic pressure

TABLE 10.1: Selected units of pressure

Unit	Value in [Pa]
1 bar	10^5
1 mbar	100
1 kp/m^2	9,80665
1 atm (physical atmosphere)	101325
1 at (technical atmosphere)	980665
1 Torr (1 mm Hg)	133,322
1 1 mm H2O	9,80665
1 PSI	6894,757293

The unit of pressure in SI system is 1 pascal (1 Pa) = 1 $[\text{kg·m}^{-1}\text{·s}^{-2}]$. Other units are still in use. Some selected ones are shown in table 10.1.

10.1   Calibration pressure gauges

Calibration pressure gauges use the definition of pressure as their principle. They measure either force on a given area or hydrostatic pressure. Calibration instruments are then used to calibrate all other pressure gauges, such as deformation gauges.

10.1.1   Bell-type pressure gauges

The bell-type pressure gauge is used for very fine measurement of low pressures. It is used for example in labs for gauge calibration. Its principle is shown in figure 10.2. Its basic component is a tank filled with a liquid with known density, for example water. An inverted bell is partially submerged in the liquid. The measured pressure is lead under the bell.

Two forces act on the bell. An external force F_e caused by lift and internal force F_i caused by gas pressure.

The internal force is

$$F_i = p \cdot A_1 \tag{10.4}$$

The external force is

$$F_e = (A_2 - A_1) \cdot l \cdot \rho \cdot g \tag{10.5}$$

In equilibrium the height l is read on the scale. The height is a function of the measured pressure p.

$$F_e = F_i \tag{10.6}$$

$$p = \frac{F_e}{A_1} = \frac{(A_2 - A_1)}{A_1} \cdot l \cdot \rho \cdot g = k \cdot l \tag{10.7}$$

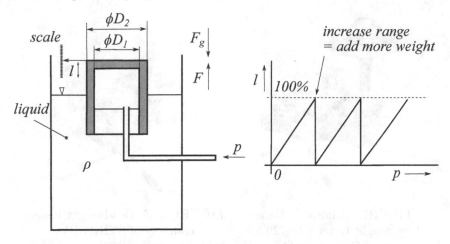

FIGURE 10.2: Bell-type pressure gauge

This manometer type can be used for small pressure ranges with high accuracy. The typical range is around 1000 Pa when water is used. The range can be extended by adding additional weights on the top side of the bell. Then the range extends as shown on the static characteristic in figure 10.2. The range extension is possible until some liquid remains under the bell. The bell-type manometer is a very accurate instrument. The typical accuracy is around 0.02% from full scale, typical resolution 0.01 Pa.

10.1.2 Deadweight testers

This manometer uses the definition of pressure, that of force acting on a known area. It is shown in figure 10.3. The force F_2 acts on a known piston area A against force F_1 caused by weight. In steady state $F_2 = F_1 = F$.

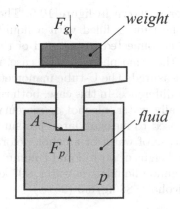

FIGURE 10.3: Deadweight tester

FIGURE 10.4: Single-Range   FIGURE 10.5: Deadweight tester,
Deadweight tester 1 to 120 bar,   courtesy of STIKO BV,
courtesy of Budenberg Gauge Co. Netherlands [51], with permissions
Ltd., UK [50], with permission

The measured pressure p is

$$p = \frac{F}{A} \tag{10.8}$$

The deadweight tester is used to calibrate other pressure sensors, for example, deformation types. It is used for high pressures, typically in range 1 to 2000 bar (100 kPa to 200 MPa). The typical error is about 0.015% of measured value. The price is around 8000 euros or more. Examples of deadweight testers are shown in figures 10.4 and 10.5.

10.1.3   U-tube manometers

A U-tube manometer is shown in figure 10.6. The transparent tube has the form of letter U. The tube is filled with a liquid of known density ρ_2. The measured pressure is connected to one end of the tube. The other end is usually left open into the atmosphere. In this case, relative pressure to the barometric pressure is measured. The U-tube manometer can however be used also to measure pressure difference. In this case, both tube ends are connected. The used liquid is typically water, alcohol or mercury. Since mercury has a large density, it can be used to measure large pressures. It is, however, toxic in larger quantities. In case of water or alcohol, colorants are usually added to simplify reading. The height of both liquid columns is read from the scale. Densities of the most common liquids are: water: 997 kg/m3 (25°C); mercury: 13534 kg/m3 (25°C); alcohol: 786 kg/m3 (25°C)

$$h = h_1 + h_2 \tag{10.9}$$

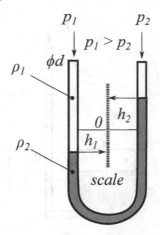

FIGURE 10.6: U-tube manometer

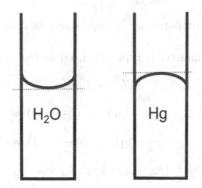

FIGURE 10.7: Correct reading from the scale

$$\Delta p = (\rho_2 - \rho_1) \cdot g \cdot h = k \cdot h \qquad (10.10)$$

where ρ_1 is the density of measured gas, and ρ_2 is the liquid density (water, Hg, alcohol,...)

The correct reading depends on the liquid type, on its surface tension, see figure 10.7. The parallax error also has to be eliminated. The observer's eye has to look in a perpendicular direction at the scale.

10.1.4 Well-type manometers (barometers)

The well-type manometer is shown in figure 10.8. It is composed of a well (tank) and a tube. The well is filled with a liquid with known density ρ_2. The measured pressure p_2, is the gas pressure, with density ρ_1 (usually air). For a constant and known tube diameter and constant liquid volume, the height h is proportional to the pressure difference $p_2 - p_1$. The change of height h_1

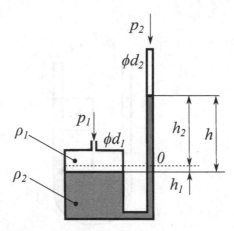

FIGURE 10.8: Principle of well-type manometer

is usually negligible compared to h_2. d_1 is significantly larger than d_2, and therefore $\Delta p = k \cdot h_2$.

The well-type manometer is typically used as barometer.

$$A_1 h_1 = A_2 h_2 \tag{10.11}$$

$$\Delta p = (\rho_2 - \rho_1) \cdot g \cdot (h_1 + h_2) =$$
$$= (\rho_2 - \rho_1) \cdot g \cdot h(\frac{A_2}{A_1} + 1) = \tag{10.12}$$
$$= (\rho_2 - \rho_1) \cdot g \cdot e \cdot h_2 = k \cdot h_2$$

where e is the conversion factor

$$e = \left(\frac{A_2}{A_1} + 1 \right) \tag{10.13}$$

The maximal range is limited by the practical size of the instrument. Typical ranges are 0 to 30 kPa (max. 3000 mm H20), accuracy 0.05 to 1%.

10.1.5   Inclined tube manometers

The inclined tube manometer is very similar to the well-type manometer. The sensitivity is increased by inclining the tube, however the range is decreased. Its principle is shown in figure 10.9, a photograph is shown in figure 10.10. The typical range is 0 to 1500 Pa (max. 1500 mm H20), accuracy around 1 Pa. The measured pressure Δp is

$$\Delta p = (\rho_2 - \rho_1) \cdot g \cdot (h_1 + h_2) =$$
$$= (\rho_2 - \rho_1) \cdot g \cdot \left(\frac{A_2}{A_1} + \sin \alpha \right) \cdot l = (\rho_2 - \rho_1) \cdot g \cdot e \cdot l = k \cdot l \tag{10.14}$$

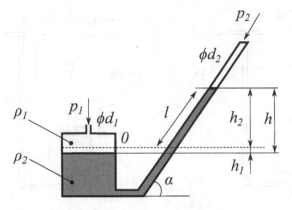

FIGURE 10.9: Inclined tube manometer

FIGURE 10.10: Inclined tube manometer—photo author

10.1.6   Compensation vacuum meters (McLeod)

This instrument is used for very small pressures, for example, for absolute pressures in range approximately 10^{-3} to 10^3 Pa. Its principle is shown in figure 10.11, and a photograph is shown in figure 10.12. The measurement is done in two steps.

In step 1 the movable tank is moved down, and the whole system is filled with the measured gas with absolute pressure p_a. The filled volume is V.

In step 2 the tank is moved up. The gas is compressed with a liquid. Assuming slow (isothermic) compression

$$p \cdot V = const. \tag{10.15}$$

The gas is compressed. The gas is enclosed into a known volume v. The tank is lifted until a mark on the fixed tank is reached. This gives the known volume v.

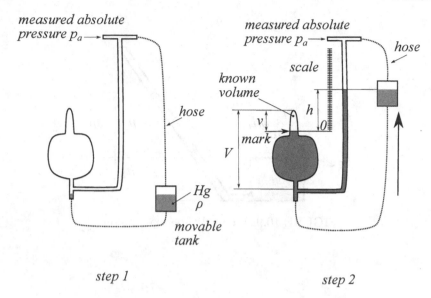

step 1   step 2

FIGURE 10.11: McLeod vacuum meter

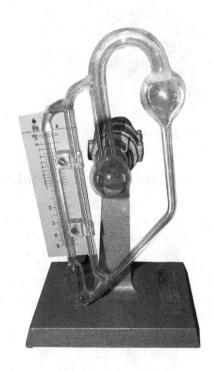

FIGURE 10.12: McLeod vacuum meter, courtesy of [52], licence CC BY 2.5

Under those assumptions

$$p_a \cdot V = (p_a + \rho \cdot g \cdot h) \cdot v \qquad (10.16)$$

If the measured pressure was larger, also a larger lift of the tank is required to reach the volume V. The measured pressure is therefore proportional to the height h

$$p_a = \frac{v}{V - v} \rho \cdot g \cdot h = e \cdot \rho \cdot g \cdot h = k \cdot h \qquad (10.17)$$

where h is the height read on the scale.

10.2 Deformation manometers

Deformation manometers use the deformation of an element to measure pressure. The element is, for example, a tube or a diaphragm. This principle is often used in industrial pressure gauges since it is very robust and reliable. On the other hand, it is not according to the pressure definition, so deformation manometers cannot be used for calibration of other gauges. Nevertheless they may have good accuracy.

10.2.1 Bourdon tubes

The Bourdon tube is shown in figure 10.13. Bourdon tubes for differential pressure are shown in figures 10.14 and 10.15. The tube is flat and wound

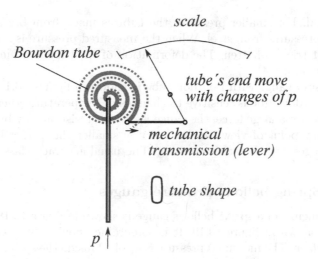

FIGURE 10.13: Principle of Bourdon-tube pressure gauge

FIGURE 10.14: Bourdon tubes for differential pressure—photo author

FIGURE 10.15: Bourdon tubes for differential pressure—photo author

FIGURE 10.16: Pressure gauge with Bourdon tube—photo author

FIGURE 10.17: Pressure gauge with Bourdon tube—photo author

into a spiral. For smaller pressures the tube is made from brass or bronze, for larger pressures from steel. When the measured pressure is applied inside the tube, it tries to flatten. The deformation of the tube is a function of the applied pressure.

A pressure gauge with Bourdon tube is shown in figures 10.16 and 10.17.

The Bourdon tube is sensitive to changes of temperature (thermal expansion). It creates a large force, the gauge is robust, resistant to vibration. From the accuracy point of view, the accuracy is smaller than other instruments, such as gauges with capacitive sensing. The usual accuracy class is > 1.

10.2.2   Spring bellows pressure gauges

The principle of a spring bellows gauge is shown in figure 10.18; the spring bellows is shown in figure 10.19. It is a metallic "box" (pipe) with a spring shaped surface. The measured pressure is applied inside the spring bellows. By increasing the pressure the spring bellows expands. A spring or the stiffness of the spring bellows itself is acting against the expansion force.

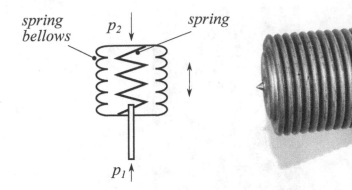

spring bellows p_2 spring

p_1

FIGURE 10.18: Principle of
spring bellows pressure gauge

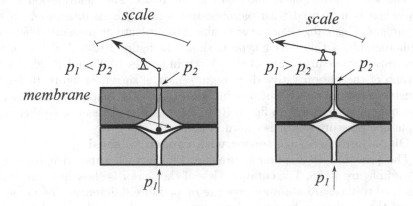

FIGURE 10.19: Spring
bellows—photo author

The expansion of the spring bellows is proportional to the applied pressure. The end of the spring bellows is connected to a needle. The change of position can also be sensed with any position sensor so the gauge may have an electric output as well. The achieved accuracy is approximately ± 2% of full scale.

10.2.3 Diaphragm gauges

The deformation element is a diaphragm as shown in figure 10.20. The deformation is proportional to the measured pressure. It is being transferred mechanically to a needle or sensed with a position sensor. Capacitive sensing or strain gauges are very often used to sense the position of the diaphragm.

scale   scale

$p_1 < p_2$ p_2 $p_1 > p_2$ p_2

membrane

p_1 p_1

FIGURE 10.20: Principe of diaphragm gauge

FIGURE 10.21: Diaphragm gauge—photo author

The material of the diaphragm can be plastics or metal based on the required pressure range. The plastic diaphragm does not have a sufficient stiffness on its own; a spring has to be used. The spring is not necessary for metallic diaphragms.

Industrial pressure gauges also use silicon diaphragms. The deformation sensor (strain gauge) can be directly integrated on the diaphragm including electronics and other required sensors, such as temperature sensors. A diaphragm gauge is shown in figure 10.21. Figure 10.22 shows a microscope photograph of a combined temperature, pressure (diaphragm), acceleration and IR thermometer sensor.

Diaphragm sensor with piezoresistors

The sensor principle is shown in figure 10.23. The deformation of the diaphragm is measured with piezoresistive strain gauges integrated on the diaphragm. This gauge can measure absolute pressure or pressure difference. A photograph of this sensor type is shown in figure 10.24. A detail of the piezoresistive pressure sensor is also shown in figures 10.25 and 10.26. A photograph of the piezoresistive diaphragm sensor is shown in figure 10.27. An industrial pressure sensor is shown in figure 10.29. The diaphragm is shown in a microscope photograph in figure 10.30. The sensing element of a differential diaphragm pressure sensor is shown in figure 10.34.

Diaphragm pressure sensor with capacitive sensing

The principle of diaphragm pressure sensor with capacitive sensing is shown in figure 10.28. The cutaway view of the sensor is show in figure 10.32. It is used to measure absolute pressure or pressure difference. The deformation of the diaphragm by applied pressure is sensed as a change of capacity. Capacitive pressure sensors are shown in figure 10.33. Industrial examples of pressure sensors are shown in figures 10.35, 10.36, 10.37 and 10.38.

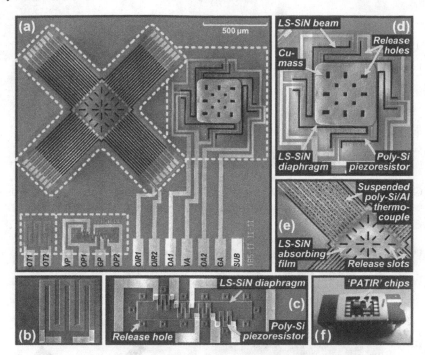

FIGURE 10.22: Combined temperature, pressure (diaphragm), acceleration, IR thermometer. Photos showing the prototype PATIR composite sensor: (a) and (f) full views, (b) thermistor, (c) piezoresistive absolute-pressure sensor, (d) piezoresistive accelerometer, (e) thermoelectric infrared detector. Courtesy of [53], license CC BY 3.0

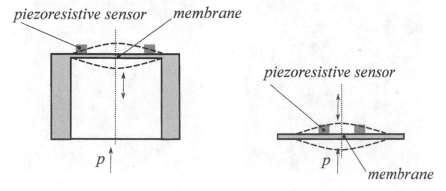

FIGURE 10.23: Principle of diaphragm sensor with piezoresistors

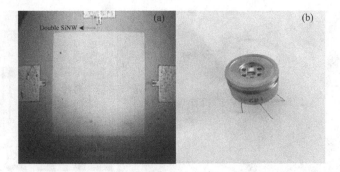

FIGURE 10.24: Piezoresistive pressure sensor, diaphragm on the left, sensor on the right; courtesy of [54], license CC BY 4.0

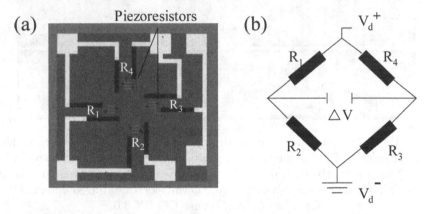

FIGURE 10.25: Piezoresistive pressure sensor, diaphragm on the left, Wheatstone bridge on the right; courtesy of [55], license CC BY 3.0

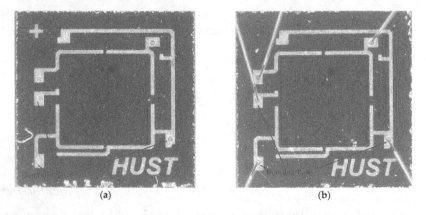

FIGURE 10.26: Piezoresistive pressure sensor, with protective layer on the left, without protective layer on the right; courtesy of [56], license CC BY 4.0

FIGURE 10.27: Piezoresistive pressure sensor, courtesy of [57], license CC BY 4.0

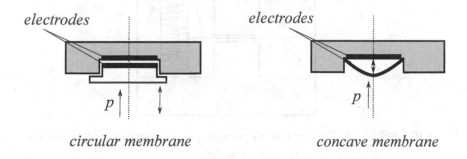

FIGURE 10.28: Principle of diaphragm pressure sensor with capacitive sensing

FIGURE 10.29: Pressure sensor with capacitive sensing—photo author

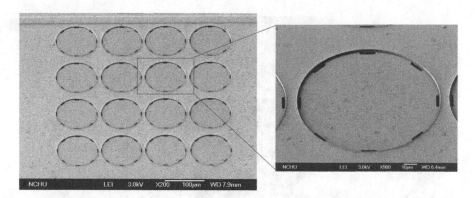

FIGURE 10.30: Diaphragm pressure sensors, courtesy of [58], license CC BY
3.0

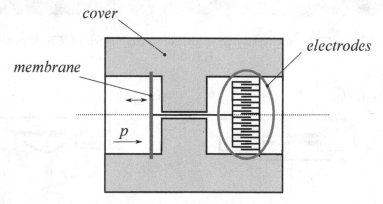

FIGURE 10.31: Principle of diaphragm pressure sensor with capacitive
sensing

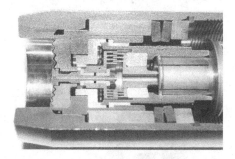

FIGURE 10.32: Cutaway view of a diaphragm pressure sensor with
capacitive sensing—diaphragm on the left, electrodes of capacitive sensor in
the middle—photo author

FIGURE 10.33: Capacitive pressure sensors—photo author

FIGURE 10.34: Sensing element of a differential pressure sensor—photo author

FIGURE 10.35: Capacitive pressure sensors—photo author

FIGURE 10.36: Capacitive pressure sensors—photo author

FIGURE 10.37: Capacitive
pressure sensors—photo author

FIGURE 10.38: Capacitive
pressure sensors—photo author

10.3   Bolometric vacuum meters—PIRANI

The principle is shown in figure 10.39, a photograph in figure 10.40 and
10.41. It is used to measure small absolute pressures (10^{-4} to 100 Pa). It uses the
heat conductivity of gasses. It is composed from a resistive bridge with two platinum temperature sensors. One sensor (A2) measures in a reference pressure,
the other (A1) the measured pressure. Both sensors are heated with current.
When pressure p_1 changes, the heat conductivity changes as well. The bridge is
unbalanced, its output voltage V_{out} is a function of the measured pressure.

10.4   Ionization pressure sensors

The principle of a ionization pressure sensor is shown in figure 10.42. Ionization pressure sensors are suitable for very small pressures, approximately

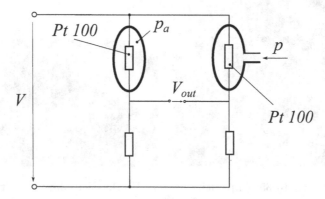

FIGURE 10.39: Principle of bolometric vacuum meter

FIGURE 10.40: Pirani Gauge Heads and Leads PV×3 Pirani Gaugeheads for use with AML Ion Gauge Controllers—courtesy of Arun Microelectronics Ltd, UK [59], with permission

FIGURE 10.41: Pirani vacuum meter—courtesy of Kurt J. Lesker Company, USA [60], with permission

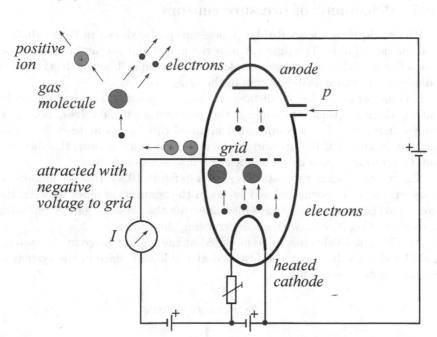

FIGURE 10.42: Ionization pressure sensor

$5 \cdot 10^{-9}$ Pa. The principle is shown in figure 10.31. It is based on the changes of conductivity of a ionized gas with changes of pressure. The main component is a heated electrode, heated to few hundred °C. A small current is created between the anode and cathode with an applied voltage. The current is a flow of electrons. The electrons hit the gas molecules and ionize the gas. The positive ions are attracted to the grid. A small current proportional to the gas pressure is created.

The grid is placed between the anode and cathode. A negative voltage is applied between the grid and the cathode; the positive ions are therefore attracted to the negative grid.

The current between the grid and the cathode is a function of gas pressure. The higher the pressure, the higher the current.

The grid current is

$$I_g = k \cdot I_a \cdot p_1 \tag{10.18}$$

Where k is a constant given by sensor construction, I_a is the current between the anode and cathode and p_1 is the measured current.

10.5   Placement of pressure sensors

Pipes, liquids, clean fluids: A pressure probe shown in figure 10.43 is used for clean fluids. The pipe between the gauge and the probe is mounted with a flange or directly welded or soldered to the pipe. The hole has typically a diameter of approximately 1 mm in the pipe wall.

It is important to keep a distance as large as possible between the probe and all elements that may change the pressure such as valves, bendings, pumps, flaps, etc. The recommended minimal distance is at least 10× pipe diameter or larger. If the pressure probe is used in a flow sensor, this distance may be up to 60× pipe diameter—see the flow sensors chapter.

The recommended installation is shown in figure 10.44. It is recommended to use special fully opened/closed valves in the connection pipes between the probe and the gauge. This allows disconnecting the pressure gauge, replacing it, re-calibrating it, etc., without interrupting the process.

An air release valve has to be mounted at the highest place in the system. A sludge drain valve has to be installed at the lowest place in the system to remove eventual sediments.

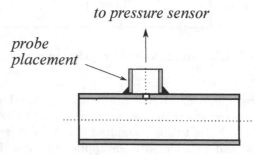

FIGURE 10.43: Pressure probe for pipes with a clean fluid

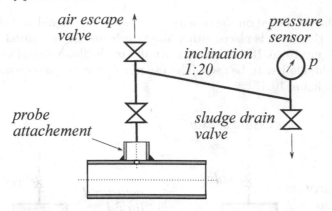

FIGURE 10.44: Recommended pressure sensor placement, including air escape valve and sludge drain valve

The connection pipes between the gauge and the probe typically have a diameter of 6 to 10 mm. Correct inclination is necessary, about 1:20 towards the sludge drain valve from the air release valve.

Hot vapors In principle a similar arrangement as for clean fluids. However condensation loops are required to protect the gauge from high temperatures. The arrangement is shown in figure 10.45.

Aggressive fluids The arrangement for aggressive fluids is shown in figure 10.46. Separation is needed to protect the gauge. Separation liquid can be, for example, water or oil. The separation liquid does not have to mix with the

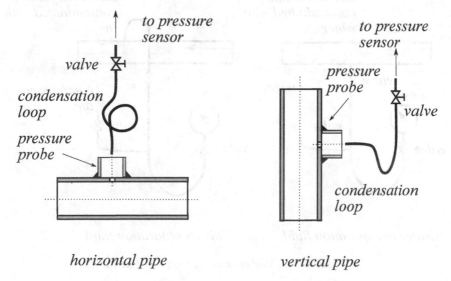

FIGURE 10.45: Vapor condensation loops

measured liquid. Based on the density of the measured liquid and the separation liquid, the gauge is placed either above or below the measured liquid—as shown in figure 10.46. If the density is similar or the fluids would mix, a separation diaphragm has to be used. The use of air or water as separation fluids is shown in figure 10.47.

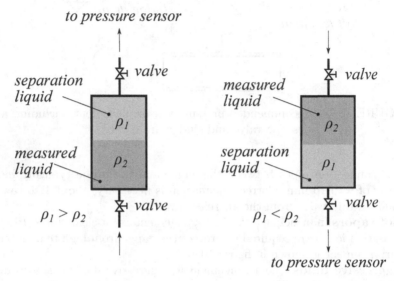

FIGURE 10.46: Use of separation liquids—arrangement based on density

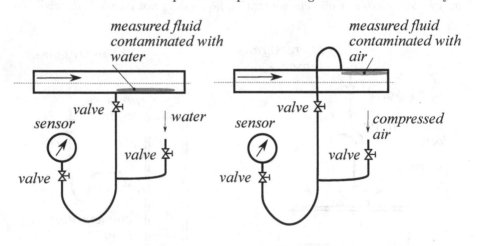

FIGURE 10.47: Water or air as separation fluids

FIGURE 10.48: Pressure sensing   FIGURE 10.49: Pressure sensing,
in beer manufacturing—photo   sight glasses, food fittings
author

Industrial examples of pressure sensors are shown in figure 10.48, where pressure sensing in beer manufacturing is shown. Figure 10.49 shows pressure sensing, sight glasses and food fittings.

11

Humidity

Absolute humidity is defined as mass of water vapor m in a unit of volume V [61]

$$\phi' = m/V \quad \left[\text{kg/m}^3\right] = [\text{kg}] / [\text{m}^3] \tag{11.1}$$

Relative humidity is defined as the ratio of current humidity to the maximal possible humidity (saturated gas).

$$\phi \doteq \frac{p_v}{p_v''} \tag{11.2}$$

where p_v is the partial pressure of water vapor of unsaturated air and p_v'' is the partial pressure of water vapour of saturated air.

The equation is approximate, as wet air is not an ideal gas.

Relative humidity depends on temperature. When temperature increases, the relative humidity decreases (for a constant absolute humidity), when the temperature decreases the relative humidity increases, until dew point is reached.

Dew point temperature is the temperature at which the air is saturated to relative humidity 100%. Decreasing the temperature below the dew point temperature leads to condensation.

The dew point temperature t_d can be calculated as follows [62]

$$t_d = \frac{B_1[\ln(\frac{RH}{100}) + \frac{A_1 t}{B_1+t}]}{A_1 - \ln(\frac{RH}{100}) - \frac{A_1 t}{B_1+t}} \tag{11.3}$$

where $B_1 = 243,04°C$, $A_1 = 17,625$, RH (%) is relative humidity and t is air temperature.

With a limited accuracy, also the equation [62] for dew point temperature t_d can be used

$$t_d \approx t - (\frac{100 - RH}{5}) \tag{11.4}$$

and for relative humidity RH

$$RH \approx 100 - 5(t - t_d) \tag{11.5}$$

The shown equations have a good accuracy only for wet air with RH > 50%.

11.1   Dew point hygrometers

The principle of the dew-point hygrometer is shown in figure 11.1. It is based on condensation on a surface cooled below the dew point temperature.

The main component is a mirror. The amount of reflected light from the mirror is measured. When the mirror temperature is below the dew point temperature, water droplets form on the mirror, and the intensity of the reflection decreases. The source of light is an LED. The absolute intensity is not of interest. Rather the drop of intensity is measured by comparing with a reference beam from the LED directly to a detector, without the mirror. The reference beam is directed directly from the LED to the detector, the second beam from the LED -> mirror -> detector. The signal is amplified in an amplifier. The control block controls the heating and cooling of the mirror, above and below the dew point temperature. The temperature waveform is shown in figure 11.2.

The dew-point hygrometer works in the following cycle [63]:

1. The mirror is quickly cooled approximately 1,5°C above the last-known dew point temperature.

2. The cooling is slowed down, the temperature is slowly decreasing to the dew point temperate. The slow decrease assures uniform dew on the mirror.

3. When the dew covers the mirror, the cooling is turned off and the heating is turned on. The mirror is heated with electric current above the dew point temperature, a few degrees higher than ambient temperature.

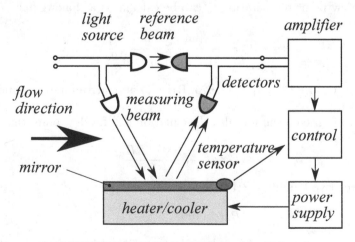

FIGURE 11.1: Dew-point hygrometer

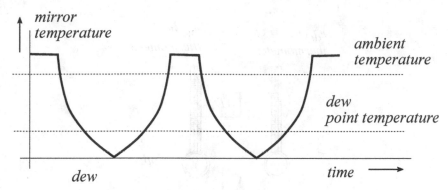

FIGURE 11.2: Temperature waveform

4. The mirror temperature is kept constant for a short time. The whole cycle then repeats.

The whole cycle takes about 35 s [64].

The cooling and heating can be done for example with a Peltier couple. The cooling and heating can be controlled with the direction of electrical current.

The mirror temperature is measured with a temperature sensor, for example with a Pt100 RTD.

The dew-point hygrometer does not change the absolute air humidity; no additional water vapor is added to the measured gas.

The detection of dew does not have to be done only optically; capacitive sensing is used as well [65].

The typical accuracy for dew point temperature is around $\pm 0,2°C$,

The dew-point hygrometer is one of the most accurate hygrometers available.

11.2 Psychrometers

A psychrometer is a very simple device used to measure relative humidity. It uses the dependency of relative humidity, temperature and temperature of saturated air. The psychrometer is composed from two thermometers or temperature sensors—wet and dry thermometer—as shown in figure 11.3 and in the photograph in figure 11.4.

The dry thermometer measures the temperature of the air where humidity is measured. The second thermometer measures the temperature of saturated air, saturated to relative humidity 100%. The increase of humidity is achieved with a wet cloth soaked with water. The cloth is wrapped around the wet thermometer.

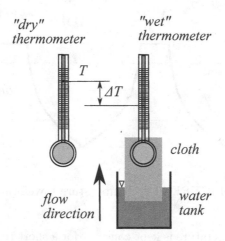

FIGURE 11.3: Principle of psychrometer

FIGURE 11.4: Psychrometer,
here with Hg
thermometers—photo author

From the psychrometric chart, in figure 11.5, or the psychrometric equation, the relative humidity is found out.

The psychrometric equation is [61]

$$p_v = p_v'' - A \cdot p \cdot (t_A - t_{WB}) \qquad (11.6)$$

where A is the psychrometric constant ($622 \cdot 10^{-6} K^{-1}$), t_A is the dry thermometer temperature and t_{WB} is the wet thermometer temperature

This principle works with temperature sensors as well. The temperature sensor can be for example an RTD or a thermistor.

The air flow is provided with a fan. For an aspiration psychrometer, the flow speed should be at least 2,5 m/s. The achievable accuracy for RH is around ±3%.

The psychrometer adds a small amount of humidity to the measured gas. It is therefore not suitable for small, enclosed spaces where it would increase humidity.

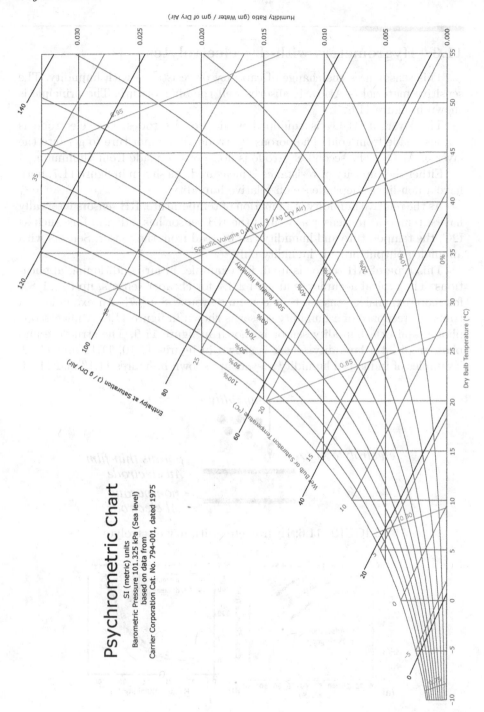

FIGURE 11.5: Psychrometric chart SeaLevel SI units, courtesy of [66], license CC BY 3.0

11.3   Hygrometers with dry electrolytes

This sensor uses the change of capacity or resistance with humidity. The sensitive material is Al_2O_3. It absorbs well the air moisture. The principle is shown in figure 11.6.

The layer of Al_2O_3 is equipped with two electrodes. One electrode is porous, made from gold. The porous electrode allows moisture to get into the layer of Al_2O_3. The second electrode is non-porous, made from aluminum.

Either the capacity or resistance is measured. As shown in figure 11.7, both have a non-linear dependence on relative humidity.

As the relative humidity is a function of temperature, RH sensors internally have a temperature sensor as well. If the RH sensor has a digital bus, such as I2C, the temperature and humidity can be read from the sensor. Sensors with an analog output provide humidity only.

This type of RH sensor is used, for example, in air conditioning applications. The typical accuracy is around $\pm 3\%$ for cheaper sensors, up to $\pm 1,8\%$ for more expensive sensors, time constant around 8 s [67]. An example of a transient response of a humidity sensor is shown in figure 11.8. A microscope photograph of a humidity sensor is shown in figure 11.9. The structure and photograph of a humidity sensor is shown in figures 11.10, 11.11 and 11.12. Examples of industrial humidity sensors are shown in figures 11.13 and 11.14.

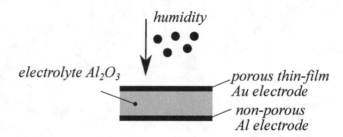

FIGURE 11.6: Hygrometer with dry electrolyte

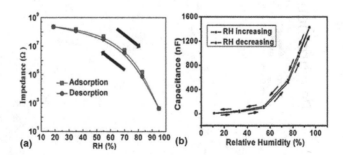

FIGURE 11.7: Dependence of impedance and capacity of RH (visible hysteresis) for a humidity sensor, courtesy of [68], license CC BY 3.0

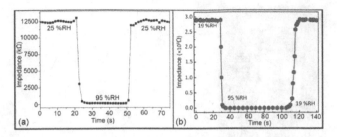

FIGURE 11.8: Transient response of a humidity sensor for various materials, a) KCl doped ZnO nanofiber, b) ZnO, courtesy of [68], license CC BY 3.0

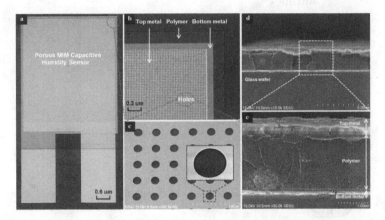

FIGURE 11.9: a) Capacitive humidity sensor, b) detail, c) detail of porous structure, d) cut through the sensor, courtesy of [69], license CC BY 4.0

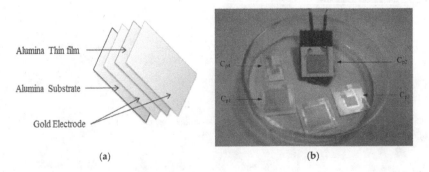

FIGURE 11.10: Capacitive humidity sensor, courtesy of [70], license CC BY 4.0

FIGURE 11.11: Humidity sensor, removed plastic cover—photo author

FIGURE 11.12: Humidity sensors—photo author

FIGURE 11.13: Temperature and humidity probe—photo author

FIGURE 11.14: Industrial humidity sensor with 4 to 20 mA current output—photo author

12

Flow

Flow can be defined in two ways—mass or volume flow.

Volume flow—volume of material passing trough an area in a unit of time

$$Q_V = \frac{V}{t} \quad (m^3 \cdot s^{-1}) \tag{12.1}$$

Mass flow—mass of material passing through an area in a unit of time

$$Q_m = \frac{m}{t} \quad (kg \cdot s^{-1}) \tag{12.2}$$

For incompressible materials (liquids or solids)

$$m = \rho \cdot V \quad (kg; kg \cdot m^3; m^3) \tag{12.3}$$

For compressible materials—gasses—the flow can be recalculated to standard conditions (p = 1 bar, T = 273 K)

$$\frac{p_0 \cdot V_0}{T_0} = \frac{p \cdot V}{T} \tag{12.4}$$

Mass or volume flowmeters are based on definition 12.1 or 12.2. They are used to calibrate other flowmeters based on different principles.

Industrial flowmeters often measure the flow velocity in a given area. Since this is not according to the flow definition, those flowmeters are not used for calibration. Nevertheless they can still have a good accuracy.

When the flow velocity (the flow velocity profile) is known, the mass flow Q_m or the volume flow Q_v can be calculated from the known cross section A

$$Q_V = v \cdot A \tag{12.5}$$

$$Q_m = A \cdot v \cdot \rho \tag{12.6}$$

where v is the mean flow speed [m/s], A is area [m^2] and ρ is density [kg/m^3].

The used principle also depends on the properties of the liquid and type of flow. For example for pipes, the type of flow is given by the Reynolds number. Types of flow for different values of the Reynolds number are shown in figure 12.1. An example of an industrial flowmeter is shown in figure 12.2.

$$Re = \frac{v \cdot D \cdot \rho}{\eta} = \frac{v \cdot D}{v} \tag{12.7}$$

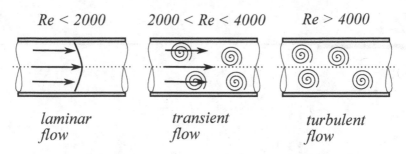

laminar *transient* *turbulent*
flow *flow* *flow*

FIGURE 12.1: Types of flow and Reynolds number

FIGURE 12.2: Sensor for pH (right) and flow (left)—photo author

where v is the mean flow speed [m/s], D is pipe diameter [m], ρ is density [kg/m^3], η is dynamic viscosity [Pa.s = kg/(m.s)] and ν is kinematic viscosity [m^2/s] ($\nu = \eta/\rho$).

12.1 Restriction flowmeters

This type of flowmeter is based on an artificial restriction in the flow. The pressure difference on the restriction is then measured. The restriction can be an orifice plate, a nozzle, Venturi tube, etc.

The measured pressure difference is a function of flow. Typically the dependence is non-linear, square root dependence.

12.1.1 Orifice plates

The orifice plate is a mechanically simple device. The principle is shown in figure 12.3, where the pressure distribution is shown as well. A photograph of an orifice plate is shown in figure 12.4. It is a flat disk with a opening with a defined diameter. The orifice plate is inserted into the pipe where the flow should be measured. In front of the orifice plate, the pipe diameter is D, the flow velocity is v_1. As the mass is constant, the same mass has to pass also through the opening in the orifice plate, where the diameter is d and the flow velocity is v_2. Since $d < D$, v_2 has to be greater than v_1.

According to the Bernoulli equation, the pressure decreases after the orifice plate. The pressure difference is a function of flow.

A permanent pressure drop occurs on the orifice plate. The leading edge of the orifice plate has to be sharp, the trailing edge has to be chamfered with a defined angle.

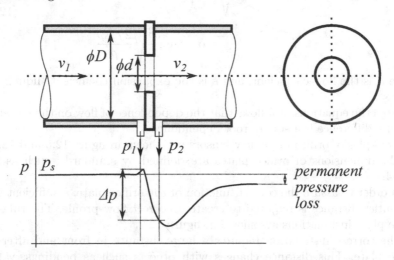

FIGURE 12.3: Orifice plate and pressure drop

FIGURE 12.4: Orifice
plate—photo author

For 1D flow of a incompressible fluid

$$Q = A_1 \cdot v_1 = A_2 \cdot v_2 \qquad (12.8)$$

$$A_1 = \frac{\pi D^2}{4} \qquad (12.9)$$

$$A_2 = \frac{\pi d^2}{4} \qquad (12.10)$$

$$A_1^2 \cdot v_1^2 = A_2^2 \cdot v_2^2 \qquad\qquad (12.11)$$

$$v_1^2 = \left(\frac{A_2}{A_1}\right)^2 \cdot v_2^2 \qquad\qquad (12.12)$$

where A_1, A_2 are pipe and orifice plate areas, and v_1, v_2 is the flow velocity in the pipe and orifice plate

From a Bernoulli equation

$$p_1 - p_2 = \frac{1}{2}\rho \left(v_2^2 - v_1^2\right) \qquad\qquad (12.13)$$

The flow is

$$Q = A_2 v_2 = \frac{A_2}{\sqrt{1 - \left(\frac{A_2}{A_1}\right)}}\sqrt{\frac{2 \cdot \Delta p}{\rho}} =$$

$$= \alpha \cdot \varepsilon \cdot \frac{\pi d^2}{4}\sqrt{\frac{2 \cdot \Delta p}{\rho}} \qquad\qquad (12.14)$$

where α is the flow coefficient, and ε is the expansion coefficient (liquids $=1$, gases <1).

From the equations, it follows that the dependence of flow on the measured pressure difference is a square root dependence.

Examples of orifice plate flow sensors are shown in figure 12.5 and 12.6.

The dimensions of orifice plates are defined by standards, such as ISO 5167-4.

In order to assure the correct function of the orifice plate, a sufficient inlet and outlet distance is required to create a correct flow profile. The rules for orifice plate installations are shown in figure 12.7.

The correct installation requires a large distance in front and after the orifice plate. This distance changes with objects such as bendings, valves, pumps, etc. It also varies with the ratio between the pipe and orifice plate diameter. The minimal distance is typically given by the ratio of pipe diameter and distance.

FIGURE 12.5: Orifice plate, flow from left to the right—photo author

FIGURE 12.6: Orifice plate—photo author

TABLE 12.1: Minimal distance in front of an orifice plate—based on [71]

β	< 0,32	0,45	0,55	0,63	0,70	0,77	0,84
A	12	12	13	16	20	27	38
B	15	18	22	28	36	46	57
C	35	38	44	52	63	76	89
E	18	20	23	27	32	40	49
F	10	13	16	22	29	44	56

Compared to other flowmeters, this distance is significant, larger than 20×
pipe diameter.

The minimal distance before the orifice plate is shown in table 12.1. **The
distances in the table are given in multiples of pipe diameter.**

The ratio between the pipe and orifice plate opening diameter is

β = d (orifice plate diameter in mm) / D (pipe diameter in mm)

After the orifice plate, the recommended minimal distance is at least 5×
pipe diameter.

The pressure taps from the orifice plate are usually connected to smart
pressure sensors for pressure difference. For example, a capacitive pressure
sensor can be used. As the flow calculation requires density, temperature sensors are used in the setup as well. Typically a Pt100 RTD is used.

Orifice plate—advantages

1. Suitable for liquids, gases and vapors

2. Suitable for high pressures (up to 42 MPa), high temperatures (up to
   1000°C)

3. The smart pressure sensor can be exchanged without interrupting the process (valves are used)

4. Suitable for large range of pipe diameters

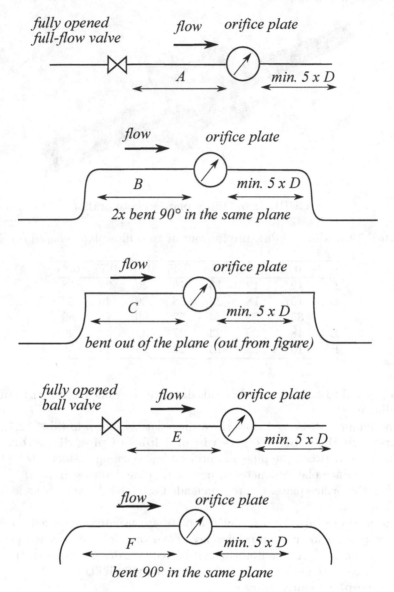

FIGURE 12.7: Installation of orifice plates

Orifice plate—limitations

1. Limited dynamic range to 4:1 or 5:1 (ratio between the maximal and minimal flow it can measure)

2. Abrasion of the orifice plate due to abrasive particles in the fluid => decrease of accuracy a lifetime of the orifice plate

3. Large required pipe length in front and after the orifice plate

4. Not suitable for small flow (<2 l/h)

5. Not suitable for fluids with large viscosity (> 500 mPa.s)

6. Does not measure directly density, viscosity, concentration as some other flowmeters (Coriolis)

7. Not suitable for abrasive fluids, such as, e.g., water–sand mixtures; abrasive materials damage the orifice plate

12.2 Rotameters

The principle of a rotameter is shown in figure 12.8, and a photograph is shown in figure 12.9. Industrial rotameters are shown in figure 12.10. It is based on a lift force acting on a float in a flowing fluid. The float rotates to assure its stability. The lifting force and gravitational forces act on the float.

The lifting force acts upwards and has the following components

$$F_A = F_p + F_d + F_f \tag{12.15}$$

F_p is caused by pressure difference

$$F_p = A_2 \cdot (p_1 - p_2) = A_2 \cdot \Delta p \tag{12.16}$$

The pressure difference Δp is virtually constant.

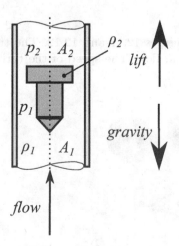

FIGURE 12.8: Principle of rotameter

FIGURE 12.9: Rotameter

FIGURE 12.10: Rotameters—photo author

F_d is the dynamic force

$$F_d = k_d \cdot A_2 \cdot p_d \qquad (12.17)$$

where k_d is a constant dependent on the float's shape

F_f is friction force

$$F_f = k_f \cdot A_f \cdot \bar{v}^n \qquad (12.18)$$

where $\bar{v}$ is the mean flow velocity, n is an exponent dependent on v and flow type, k_f is a constant dependent on the fluid viscosity, A_f is the friction area.

The measured flow is then

$$Q_v = k \cdot \alpha \left(A_1 - A_2\right) \sqrt{\frac{p_1 - p_2}{\rho_1}} \qquad (12.19)$$

12.3   Turbine flowmeters

The turbine flowmeter is shown in figures 12.11 and 12.12. The turbine inside the flowmeter is turned by the fluid flow. The speed of the turbine is proportional to flow.

FIGURE 12.11: Turbine flowmeter (here for water flow)—photo author

FIGURE 12.12: Turbine flowmeter—photo author

12.4 Wire anemometers

The principle of wire anemometers is shown in figure 12.13, with photographs in figures 12.15 and 12.14. The flowmeter is based on the temperature measurement of the wire. Three ways of measurement are used.

a) Wire powered with constant current

In this case, the wire is powered with a constant current from a current source. The temperature of the wire is measured with a temperature sensor or determined from the wire resistance. Platinum wire is used in the sensor. As the wire is cooled down with the measured flow, the wire temperature decreases, and this can be measured as a decrease it its electrical resistance. Resistance is then a function of flow velocity R = f(v).

b) Wire kept at constant temperature

The wire temperature is maintained at a constant temperature with a feedback loop. The current in the wire is adjusted to maintain a constant temperature. The current is proportional to the measured flow velocity I = g(v).

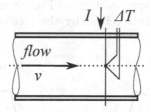

FIGURE 12.13: Principle of wire anemometer

FIGURE 12.14: Probe of a wire anemometer—photo author

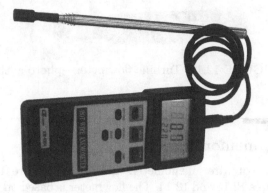

FIGURE 12.15: Wire anemometer—photo author

c) Constant power

A constant power is supplied to the wire. The wire is heated to a temperature approximately 350°C. The wire temperature is measured, for example, with a thermocouple. The wire is cooled down with the measured fluid. The wire temperature is a function of flow velocity t = h(v).

12.5   Ultrasonic flowmeters

The principle of ultrasonic flowmeters is shown in figure 12.16. They are equipped with ultrasonic transmitters and receivers. The transmitters send signals, and the signals are received by the receivers. When the ultrasonic

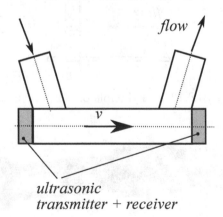

FIGURE 12.16: Principle of ultrasonic flowmeter

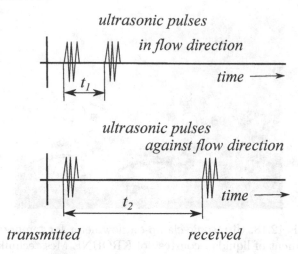

FIGURE 12.17: Timing of ultrasonic signals

signal travels upstream it is being accelerated by the flow. If it travels downstream it is being slowed down. The time difference between the upstream and downstream timing is a function of flow velocity. The sensor features at least one pair of receiver and transmitter. With more pairs the flow profile can be measured more accurately.

The ultrasonic transmitter 1 (in figure 12.16 on the left side) is received by the receiver. The propagation delay is measured. Next transmitter 2 on the right side sends a signal. It is received by receiver 2, and the time delay is measured as well. With 0 flow velocity, both times are equal. The timing of ultrasonic signals is shown in figure 12.17.

For flow direction from left to right, the ultrasonic signal traveling upstream travels faster, and the time delay is smaller. The flow is calculated from the measured flow velocity and known pipe cross section.

The ultrasonic flowmeter measures flow in both directions (for example, an orifice plate measures only in one direction).

The ultrasonic flowmeter can also be clamped externally on the pipe as shown in figure 12.18. No armatures are required, there is no pressure drop. Also aggressive liquids can be measured with the clamp-on system.

An example of an industrial inline ultrasonic flowmeter is shown in figure 12.19.

The installation of ultrasonic flowmeters in pipes is shown in figure 12.20. The ultrasonic flowmeter is sensitive to particles suspended in the fluid. It is also sensitive to bubbles. Both disperse the ultrasonic signal and cause its attenuation.

The recommended maximal content of those components is
Gases: <2% of volume
Solids: <5% of volume

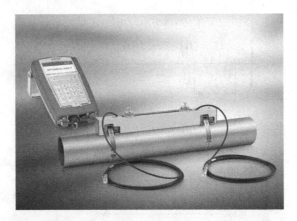

FIGURE 12.18: Ultrasonic clamp-on flowmeter for temporary flow
measurement of liquids—courtesy of KROHNE Messtechnik GmbH,
Germany [72], with permission

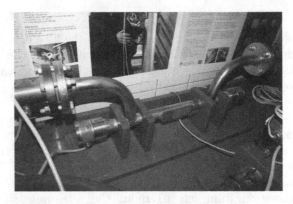

FIGURE 12.19: Ultrasonic flowmeter—photo author

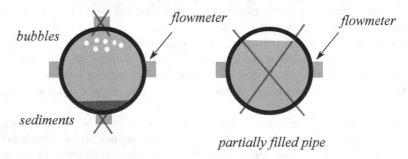

FIGURE 12.20: Installation of ultrasonic flowmeters, and bubble and
sediment removal by mounting the sensor from the pipe sides. The pipe has
to be full at all times, not partially

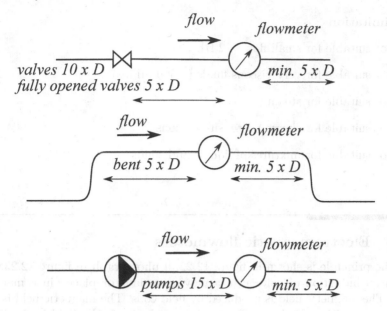

FIGURE 12.21: Ultrasonic flowmeter—correct installation

It is also necessary to have the pipe completely filled with the fluid, since the flow is calculated from flow velocity and pipe cross section.

The correct installation of an ultrasonic flowmeter with respect to valves, bendings etc. is shown in figure 12.21.

Properties of ultrasonic flowmeters

1. Suitable for clean liquids and gases

2. Accuracy approximately 2% in industrial environment when correctly installed, up to 0,5% possible

3. Suitable for large pipe diameters, up to DN 5100 (clamp on), DN 2000 (armature)

4. Temperature range approximately: −50 to +170 °C

Advantages

1. Dynamic range approximately 50:1 (for liquids)

2. Zero pressure loss for the clamp on system

3. No damage from abrasive or aggressive fluids

4. Measures both flow directions

5. Clamp-on system can be installed without interrupting the process.

6. Price virtually independent on the pipe diameter (for clamp-on system)

Limitations

1. Not suitable for small flow < 2 l/h

2. Not suitable for very viscous fluids (>500 mPa.s)

3. Not suitable for steam

4. Not suitable for slurry, paste, suspensions

5. Not suitable for mixtures of liquids and solids

12.6   Electromagnetic flowmeters

The principle is shown in figure 12.22, a photograph in figure 12.23. It is based on induction of voltage in a moving conductor placed in a magnetic field. The magnetic field is produced by field coils. The magnetic field is then distributed with pole shoes. The moving conductor is the fluid itself. Therefore the fluid has to be conductive. This flowmeter can be used only with electrically conductive fluids.

For zero flow velocity, the positive and negative ions are uniformly distributed in the fluid.The fluid is in a magnetic field. When the fluid starts to flow a force starts to act on the electrically charged particles. Positive ions move towards one side of the pipe, the negative ions to the other. An electric field is created. The electric field is measured with electrodes. The electrodes are in a perpendicular direction to the magnetic field.

The measured voltage is small, typically around 300 µV per m/s of flow velocity. The measured voltage is directly proportional to flow velocity. The flow is calculated from the measured flow velocity and known pipe cross section.

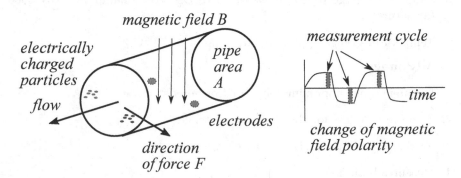

FIGURE 12.22: Principle of electromagnetic flowmeter

FIGURE 12.23: Electromagnetic flow meter—photo author

External magnetic fields need to be eliminated. This is achieved by altering the polarity of the magnetic field. The AC signal is then evaluated at the electrodes. This also eliminates electrochemical effects in the fluid.

Expected parameters

1. Suitable only for conductive fluids (conductivity > 1 μS/cm)

2. Accuracy around 0,5% (0,2%)

3. For pipe diameters up to DN 2510

4. For pressures up to 4 MPa

5. Temperature range approximately −50 to +180°C

Advantages

1. Dynamic range up to 1000:1

2. Zero pressure loss

3. Suitable for abrasive mixtures of liquids and solids, such as water–sand mixtures, pastes etc.

Limitations

1. Requires **conductive** liquid

2. Problematic when sediments occur in the pipe (flow velocity is measured)

3. Does not measure cryogenic liquids (liquid O2, Ar, N2)

4. Does not measure gases, vapors

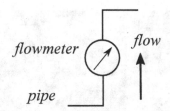

FIGURE 12.24: Installation in a vertical pipe—flow upwards so that the pipe is full at all times

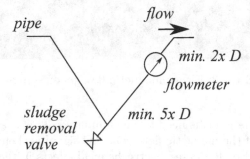

FIGURE 12.25: Installation in a horizontal pipe—the flowmeter is full at all times

Correct installation of magflow flowmeters

The correct installation has to assure that the flowmeter is full at all times. An example of installation into a horizontal pipe is shown in figure 12.25. Minimal upstream and downstream distances are also given.

For a vertical installation, the correct installation is shown in figure 12.24. The pipe has to be full at all times. The flowmeter has to be without sediments and bubbles. Both should accumulate outside the flowmeter, the sediments at the bottom, the bubbles at the top.

12.7   Coriolis flowmeters

The Coriolis flowmeter measures mass flow directly. Its principle is shown in figure 12.26, and a photograph is displayed in figure 12.27. It can independently measure density as well. It is based on the effects of the Coriolis force on a liquid in a moving tube. The tube is moving with a defined angular speed, both ends of the tube move in different directions. The movement of the tube is measured with position sensors.

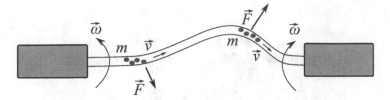

FIGURE 12.26: Principle of Coriolis flowmeter

FIGURE 12.27: Coriolis flowmeter—photo author

The Coriolis force is proportional to the angular speed ω of the pipe and flow velocity of the fluid v.

$$\vec{F_c} = 2 \cdot m \cdot (\vec{v} \, x \, \vec{\omega}) \tag{12.20}$$

The Coriolis force causes both tube ends to move differently. The ends oscillate with the same frequency but different phase shift. The phase shift is proportional to the measured flow. The position is measured with position sensors, e.g. inductive sensors with continuous output.

It is possible to measure the fluid density independently. The resonance frequency is a function of fluid's density

$$f_R = \frac{1}{2\pi} \cdot \sqrt{\frac{c}{m_{fl} + m_t}} \tag{12.21}$$

where $m_{fl} = V \cdot \rho_{fl}$ is the fluids mass in the pipe and m_t is the mass of the pipe

Both measurements can be done simultaneously but independently.

The correct installation of a Coriolis flowmeter in a pipe is shown in figure 12.28.

FIGURE 12.28: Coriolis flowmeter—courtesy of [73], license CC BY 3.0

Advantages

1. Measures directly mass flow of liquids and gases, density and temperature (typically with an embedded Pt1000)

2. Very high accuracy—typically 0,1% (až 0,05%)

3. Does not require inlet and outlet distances

4. Possible to calculate volume flow, viscosity, concentration (e.g., sugar in water, alcohol in water, etc.) assuming a single phase fluid

5. Possible to measure small flow < 2 l/h

6. Pipe diameters DN 1 up to DN 510

7. Pressures up to 40 MPa

8. Temperature range approximately −50 to +350 °C

Limitations

1. Higher pressure loss compared to clamp-on ultrasonic and magflow flowmeters

2. Optimal only with single-phase fluids, two-phase fluids have large impact on accuracy

3. High price

13

Liquid level

Liquid-level sensors can be divided into two categories, based on the output signal and detection:

1. Continuous = measure the liquid level continuously, the output signal is a number on a digital bus, analog voltage or current

2. Limit = measure if the liquid level is above or below. The output signal is a logic signal, 0 or 1.

13.1   Visual liquid-level meters

The visual liquid-level meter is the simplest form of a liquid-level sensor. It is shown in figure 13.1, and it is often used to indicate the liquid level in tanks, without any electric output.

The pipe of the visual liquid-level meter is transparent; it is connected with the main tank with a pipe. The transparent pipe is equipped with a scale.

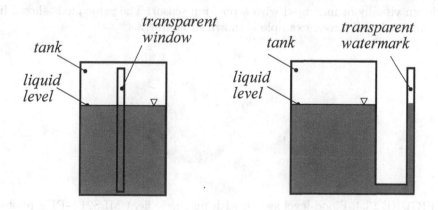

FIGURE 13.1: Visual liquid-level meter

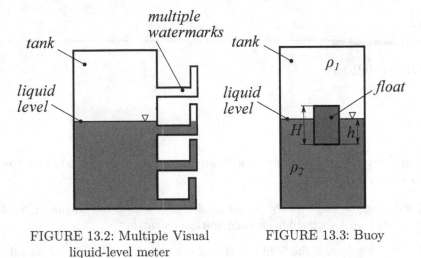

FIGURE 13.2: Multiple Visual   FIGURE 13.3: Buoy
       liquid-level meter

Due to different temperatures in the liquid-level meter and the main tank, problems can arise with thermal expansion of the liquid. The visual liquid-level meter can then be split into multiple sections as shown in figure 13.2.

13.2 Floats

The basic component is a float, lifted by buoyancy. The float's density has to be smaller that the density of the liquid. The position of the float can be shown visually or measured with a position sensor. The principle is shown in figure 13.3, and sensor example is shown in figure 13.4.

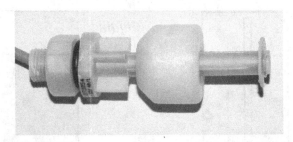

FIGURE 13.4: Flood-level switch with magnetic float MLS21A-PP—photo
author

The buoyancy is

$$F = A \cdot H \cdot \rho_B \cdot g - A \cdot h \cdot \rho_1 \cdot g - A(H-h) \cdot \rho_2 \cdot g \qquad (13.1)$$

where A is the float's area, ρ_B is float's density, g is gravitational acceleration, ρ_1 is liquid density, ρ_2 is gas density above the liquid.

13.3 Hydrostatic sensors

The sensor is a pressure sensor installed on the bottom of the tank. It measures the hydrostatic pressure of the liquid above it. It is required to know the liquid's density.

The height of the liquid column between the tank's bottom and the pressure sensor also needs to be accounted for, as shown in figure 13.5. In figure 13.5, a well-type manometer is shown as an example; any other pressure sensor type can be used. An example of a industrial hydrostatic liquid-level sensor is shown in figure 13.6.

13.4 Bubbler

The main component of a bubbler, shown in figure 13.7, is a pipe submerged in the tank. Pressurized gas is blown in the pipe, and bubbles are

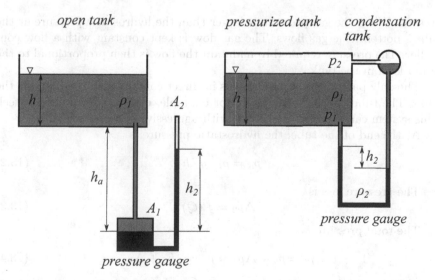

FIGURE 13.5: Hydrostatic sensor in open and closed (pressurized) tanks

FIGURE 13.6: Hydrostatic liquid-level sensor—photo author

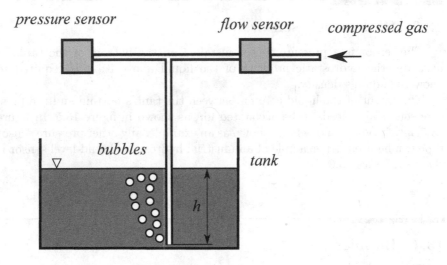

FIGURE 13.7: Bubbler

created. When the gas pressure is larger than the hydrostatic pressure at the tube's bottom, the gas flows. The gas flow is kept constant with a flow controller. The pressure required to maintain the flow is then proportional to the liquid level in the tank.

The only part of the system that is in direct contact with the liquid is the tube. The tube can be made from, for example, ceramics or stainless steel. This system can therefore work well with aggressive or viscous liquids.

At the end of the tube, the hydrostatic pressure is

$$p_h = \rho_1 \cdot g \cdot h_1 \tag{13.2}$$

The pressure loss is

$$\Delta p_2 = f(Q) \tag{13.3}$$

The total pressure is

$$p_t = p_h + \Delta p_2 = \rho_1 \cdot g \cdot h_1 + \Delta p_2 \tag{13.4}$$

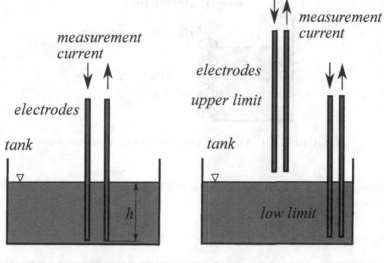

FIGURE 13.8: Continuous and limit conductivity level sensor

13.5 Electrical conductivity-level sensors

The principle is shown in figure 13.8. This sensor may be used for both continuous and limit sensing. It requires an electrically conductive liquid.

The electric conductivity of the fluid is used. For continuous sensing, two electrodes are submerged into the fluid. The current passing through the fluid is proportional to the liquid level. As the liquid level increases, the conductivity increases.

In case of limit sensing the fluid connects the two electrodes when the desired liquid level is reached. The connection is detected.

13.6 Thermal conductivity-level sensors

The principle is based on different heat conductivity when a heated element is or is not submerged into a liquid. The heated element can be, for example, a thermistor as shown in figure 13.9. This principle is limit sensing not continuous. The sensor is heated, and its temperature is measured. When the sensor is submerged into a liquid, the thermal conductivity is greater than when it is in air.

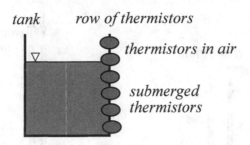

FIGURE 13.9: Array of thermistors as level sensors

The sensors can be installed in an array to get more measured liquid levels as shown in figure 13.9.

13.7 Radioisotope liquid-level meters

The principle of radioisotope liquid-level meters is shown in figure 13.10. It is based on attenuation of nuclear gamma radiation in a material placed between the source and detector. The dependence of intensity on distance is quadratic; the dependence on thickness is exponential.

The radiation source is placed on the bottom of the tank, the detector on top. The detected intensity is a function of material thickness (liquid level).

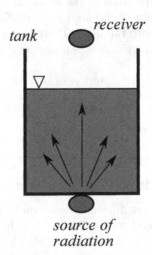

FIGURE 13.10: Radioisotope liquid-level meter

The source of nuclear radiation is Co60. The measured substance does not become radioactive. The sensor can measure fluids, powders or sludge. It is often used in the food processing industry.

The typical accuracy is $+/- 0.5\%$, range around 7 m.

13.8 Capacitive liquid-level sensing

This sensor uses the changes of capacity of a capacitor caused by changes of liquid level. Its principle is shown in figure 13.11. The dielectric material is the liquid. One electrode is a wire or a rod; the second electrode is a wire or a rod as well, or it can be the tank itself. When a conductive liquid is measured, one of the electrodes needs to be insulated.

The measured liquid needs to have a significantly different relative permittivity ε_{r1} from the gas above the liquid (typically air).

For a planar capacitor the capacity is

$$C = \varepsilon_0 \cdot \varepsilon_r \frac{A}{d} \qquad (13.5)$$

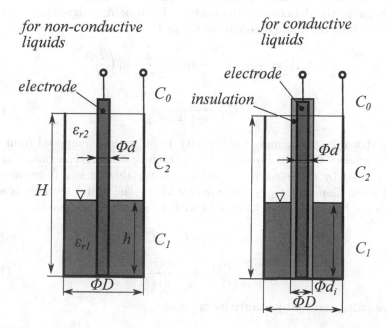

FIGURE 13.11: Principle of capacitive liquid level sensors—for conductive and non-conductive liquids

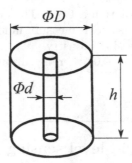

FIGURE 13.12: Cylindrical capacitor

where ε_0 is vacuum permittivity ($\varepsilon_0 \approx 8,854187817 \times 10^{-12} F \cdot m^{-1}$), ε_r is relative permittivity (material property), A is electrode area and d is electrode distance.

For a cylindrical capacitor shown in figure 13.12, the capacity is

$$E = \frac{\lambda}{2\pi \cdot \varepsilon \cdot r} \tag{13.6}$$

where λ is charge per unit distance, r is radius ($r = D/2$).

Assuming the diameter of the central electrode d is significantly smaller than the diameter of the external electrode D

$$\Delta V = \frac{\lambda}{2\pi \cdot \varepsilon} \int_d^D \frac{1}{r} dr = \frac{\lambda}{2\pi \cdot \varepsilon} \ln\left(\frac{D}{d}\right) \tag{13.7}$$

$$C = \frac{2\pi \cdot \varepsilon}{\ln\left(\frac{D}{d}\right)} \cdot h \tag{13.8}$$

As shown in the figures, the capacity is in general composed from three parts. Capacity C_0 is the capacity of the attachment and is constant. Capacity C_1 is caused by the measured liquid and is variable. It is a function of the liquid level. Capacity C_2 is the capacity above the liquid level. It is also a function of the liquid level (decreases with increasing liquid level).

$$C = C_0 + C_1 + C_2 \tag{13.9}$$

$$C = C_0 + \frac{2\pi\varepsilon_0\varepsilon_{r1}}{\ln\left(\frac{D}{d}\right)} \cdot h + \frac{2\pi\varepsilon_0\varepsilon_{r2}}{\ln\left(\frac{D}{d}\right)} \cdot (H - h) \tag{13.10}$$

Assuming constant permittivity ε_{r1} and ε_{r2}

$$C = k_1 + k_2 \cdot h \tag{13.11}$$

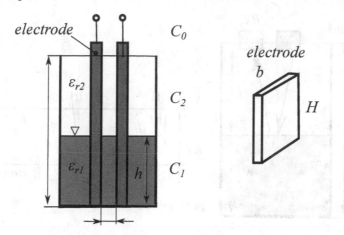

FIGURE 13.13: Planar electrodes

where

$$k_1 = C_0 + \frac{2\pi\varepsilon_0\varepsilon_{r2}}{\ln\left(\frac{D}{d}\right)} \quad k_2 = C_0 + \frac{2\pi\varepsilon_0}{\ln\left(\frac{D}{d}\right)} \cdot (\varepsilon_{r1} - \varepsilon_{r2}) \tag{13.12}$$

For planar electrodes, shown in figure 13.13, the capacity is

$$C = C_0 + C_1 + C_2 \tag{13.13}$$

$$C = C_0 + \varepsilon_0 \cdot \varepsilon_{r1} \frac{b \cdot h}{d} + \varepsilon_0 \cdot \varepsilon_{r2} \frac{b \cdot (H - h)}{d} \tag{13.14}$$

13.9 Ultrasonic liquid-level meters

The principle of ultrasonic liquid-level sensors is shown in figure 13.14; an industrial sensor is shown in figure 13.15. An ultrasonic pulse is transmitted from the transmitter. The signal is reflected from the liquid level. The reflection is received by the receiver. The time for the reflection to arrive is measured. The time is proportional to the distance from the sensor to the liquid level.

The used frequencies are in the order of kHz to few MHz, the range is approximately from 0.01 to 15 m, accuracy is typically in the order of mm.

The measured distance is

$$l = \frac{c \cdot t}{2} \tag{13.15}$$

where c is speed of sound in the gas above the liquid, and t is the measured time delay between the transmission and reflection of ultrasonic signal.

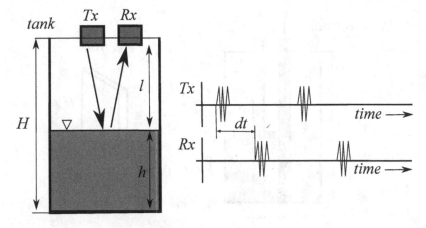

FIGURE 13.14: Principle of ultrasonic liquid-level sensor

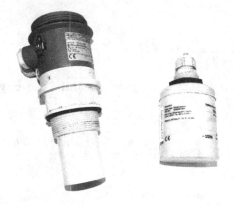

FIGURE 13.15: Ultrasonic liquid-level sensors—photo author

The measured signal is not influenced by changing properties of the measured liquid, such as density. On the other hand, it can be influenced by bubbles or foam at the surface.

13.10   Radar liquid-level meters

The sensor used the reflection of radio waves from the liquid level. Frequency bands of 6 GHz, 26 and 80 GHz are used. The signal is not influenced by vapors above the liquid. Two principles are used.

a) Time of Flight (TOF)

TOF is used in general for larger distances. The radio-wave pulse is transmitted from the transmitter. The signal is reflected from the liquid level. The reflection is received by the receiver. The time between the transmission and the reflection is measured. As the radiowave travels with speed of light, the time difference is very small, in the order of fs for small distances. For smaller distances FMCW is used.

b) Frequency-Modulated Continuous Wave (FMCW)

The principle is shown in a block diagram in figure 13.16. A continuous wave with variable frequency is transmitted. The signal is reflected from the liquid level. The reflection is received by the receiver. The received frequency is compared with the transmitted frequency. The frequency difference is proportional to the measured distance. The maximal possible distance is limited by the available frequency sweep. The achieved accuracy is in the order of mm. An example of a radar liquid-level sensor is shown in figure 13.17.

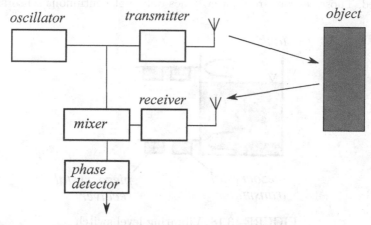

FIGURE 13.16: Principle of FMCW

FIGURE 13.17: Radar liquid-level sensor, without antenna—photo author

The measured distance is

$$l = \frac{c_0 \cdot t}{2\sqrt{\mu_r \varepsilon_r}}$$ (13.16)

where c_0 is speed of light in vacuum, μ_r and ε_r is relative permittivity and permittivity of the material above the liquid (air), respectively.

13.11 Vibrating-level switches

The principle of vibrating-level switches is shown in figure 13.18 and the photograph in figure 13.19. The main component is a mechanical resonator, a vibrating fork. Vibrations are excited with a piezocrystal. When the fork is submerged into a liquid or a solid, its resonance frequency changes. This is detected. Those sensors are limit switches only, not continuous sensors.

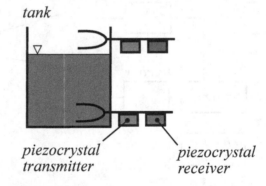

FIGURE 13.18: Vibrating-level switch

FIGURE 13.19: Vibrating-level switches—photo author

14

Example labs

This chapter presents few selected laboratory tasks as an inspiration for tutorial classes in the field of sensors.

14.1 Temperature—contact thermocouples

14.1.1 Introduction

Thermocouples are manufactured by welding of thermocouple wire with a setup shown in figure 14.1. In this task we will weld a thermocouple, make its calibration certificate and measure its steady-state response.

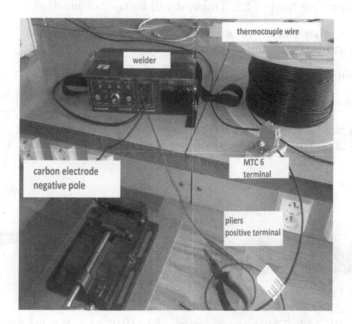

FIGURE 14.1: Photo of the task

14.1.2 Tasks

1. Weld a J-type thermocouple, length 50 cm.

2. Make a calibration certificate for the thermocouple for temperatures 30°C, 55°C and 80°C.

3. Measure steady-state response.

14.1.3 Instruments used

- Thermocouple welder TW163 with accessories

- Agilent 34970A data acquisition unit with MTC 6 terminal for J-type thermocouples

- Thermocouple wire 7960017 J JX / $2 \cdot 0{,}22$ mm^2

14.1.4 Procedure description—thermocouple welding

1. Cut off the thermocouple wire. Length 50 cm.

2. At one end remove the insulation in length 1 cm, twist the wires together—see figure 14.2. This end will be the hot junction.

3. At the other end of the wire, remove the insulation in length of 3 cm. Here will be the terminal connection.

4. Put on the protective glasses.

5. Set the welder into AUTO mode, set the energy to 30 J

6. Hold the future hot end of the thermocouple with the pliers. Press the CHARGE button on the welder and wait until the READY light turns on.

FIGURE 14.2: Thermocouple wire ready for welding   FIGURE 14.3: Welded wire at carbon electrode of the welder

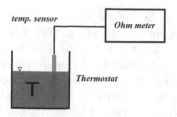

FIGURE 14.4: Correctly welded thermocouple hot end—ball at the end

FIGURE 14.5: Block diagram of calibration and to measure static response

7. Now really put on the protective glasses!

8. Put the future hot end to the carbon electrode—figure 14.3. The welder starts welding automatically after 1.5 s from the detection of connection.

9. The correctly welded hot end is shown in figure 14.4. There has to be a small ball at the end. If not, you should repeat the procedure.

10. Soak the thermocouple hot end into a protective lacquer. The lacquer dries in about 1 to 2 min. The thermocouple is now ready for calibration.

14.1.5 Procedure description—calibration and static response

1. Connect the thermocouple to the MTC 6 terminal. Take care of the correct wire overlap.

2. Submerge the thermocouple hot end into the hot water in thermostat.

3. For water temperature in range from 30°C to 80°C measure the thermocouple voltage. Change the temperature in 10°C steps. The block diagram is shown in figure 14.5.

4. Compare the measured voltage with a J-type thermocouple table.

5. Plot the steady-state response from the measured data.

14.2 Temperature—non-contact emissivity

14.2.1 Introduction

When working with non-contact thermometers the key term is emissivity ϵ. It is defined as *"the relative power of a surface to emit heat by radiation; the ratio of the radiant energy emitted by a surface to that emitted by a blackbody at the same temperature"* [74]. The user needs to be aware that a non-contact thermometer is not measuring temperature but radiated energy. In order to show temperature, the emissivity of the measured object has to be known and correctly set on the thermometer. The purpose of this experiment is to estimate emissivity of different surfaces and to estimate the error caused by the wrong setting of emissivity (or fixed—some IR thermometers don't allow to change emissivity).

14.2.2 Tasks

1. With contact thermocouple and IR thermometer with fixed emissivity 0.95, determine emissivity of different surfaces.

2. Calculate absolute and relative error of temperature readings caused by the different emissivity of the object.

3. Make a thermographic image of the Al heat sink with the thermocamera.

14.2.3 Instruments used

- Fixtures with resistors on a heatsink, Peltier element

- IR thermometer with fixed emissivity 0,95

- Multimeter Axiomet AX-18B with surface thermocouple probe

- Thermocamera FLIR i50

- Power supply

14.2.4 Procedure description

The radiation of a black body is described by the Planck's law. It relates the spectral radiance of a black body (the quantity of radiation) with its temperature and wavelength of the radiation. A blackbody emits total radiant power W_B into a surrounding hemisphere given by

$$W_B = \sigma \cdot T^4 \tag{14.1}$$

where σ is Stefan-Boltzman constant, and T is temperature in Kelvin.

Any other body can be characterized by a dimensionless parameter—emissivity ϵ

$$\epsilon = W/W_B \tag{14.2}$$

It is the fraction of black body power emitted in the surrounding hemisphere. Emissivity depends on the surface of the body and on its temperature. By definition, it is 1 for a black body. The black body is an idealized concept. Real objects do not absorb all incident energy; some part is reflected. They behave like gray bodies. Their emissivity is $\epsilon < 1$.

In order to correctly measure temperature with an IR thermometer, the emissivity of the measured object has to be known. The method used here consists in measuring the real object temperature with a contact thermometer (thermocouple in this case). The used IR thermometer has a fixed setting of emissivity 0.95. Therefore, its reading is correct only for this emissivity. The IR thermometer measures the radiated energy

$$W_{IR} = \epsilon_{0.95} \cdot \sigma \cdot T_{0.95}^4 \tag{14.3}$$

where $T_{0.95}$ is the temperature shown on the IR thermometer with fixed emissivity 0.95. The object-radiated energy is a function of its temperature T_{obj}.

$$W_{obj} = \epsilon_{obj} \cdot \sigma \cdot T_{obj}^4 \tag{14.4}$$

In order to determine object emissivity ϵ_{obj}, the object temperature T_{obj}. is measured with a contact thermometer. Then the object emissivity ϵ_{obj} can be calculated as

$$\epsilon_{obj} \cdot T_{obj}^4 = \epsilon_{0.95} \cdot T_{0.95}^4 \tag{14.5}$$

hence

$$\epsilon_{obj} = \epsilon_{0.95} \cdot T_{0.95}^4 / T_{obj}^4 \tag{14.6}$$

The temperatures have to be substituted in Kelvin.

14.3 Position—LVDT

14.3.1 Introduction

The LVDT is a linear displacement position sensor. It transfers the movement of the transformer core into the output signal. The output signal is a voltage signal; in our case the used sensor has built-in electronics to transfer the voltage to current output. Sensors with current output (4 to 20 mA) have a limit of maximal resistance that can be connected to sensors' output. The ideal load for a current output sensor is resistance zero; the maximal load is limited by the available voltage for output. When the output loop resistance is increasing, the output compensates by increasing output voltage so as to keep constant current. When the output is already on the maximal voltage, the output current starts to drop. The maximal resistance connected to the current output limits the wiring length between the sensor and gauge.

14.3.2 Tasks

1. Measure and plot the static characteristic $I = f(x)$ for load resistor $R_L = 0$ ohm.

2. For constant position ($x = 100$ mm) determine the maximum value of load resistance R_L, until which the output current remains constant. Calculate the sensitivity of the transducer.

14.3.3 Instruments used

- Fixture with LVDT Monitran MTN IE(I)S-75

- DC ampere meter

- Decade resistor R_L

14.3.4 Procedure description

The industry standard for current output is 0 to 20 mA or 4 to 20 mA. It is defined by the ANSI/ISA–50.00.01–1975 (R2002) standard and has wide industrial usage. The advantage is that the reading is not dependent on the wire length between the sensor and gauge. The limitation of wire length is imposed by the connecting wire resistance and by the maximal load resistance of the transmitter. In order to push a constant current through the loop (dependent only on the measured variable), the transmitter is adjusting its output voltage. When the output loop resistance is increasing (e.g., by changes of temperature, oxidation, etc.), the output compensates by increasing output voltage so as to keep constant current. When the output is already on the maximal voltage, the output current starts to drop.

1. Adjust the measured distance, read the LVDT output current, plot steady-state response

2. Set a constant position x, change the load resistance R_L on the decade resistor, measure the loop current. The current will stay constant until a limit value is reached. Then the output voltage cannot keep up with the resistance change to maintain constant current.

14.4 Influence of material to proximity position sensors

14.4.1 Introduction

FIGURE 14.6: Experiment proximity sensors

When selecting a position sensor one of the important parameters is the material of the object to be detected, for example, for ferromagnetic materials inductive sensors work quite well. But for non-ferromagnetic metals (Al, Cu,...), their sensing distance decreases considerably, and for non-metallic materials they can't be used. Capacitive sensors, on the other hand, detect non-metallic objects as well, but their sensing distance is also a function of the object material (permittivity). Optical sensors may have problems with transparent objects. In this task, we will examine the properties of different sensors with different materials. The setup is shown in figure 14.6.

14.4.2 Tasks

1. Find out what sensor principles work with what materials.

2. For an inductive sensor, determine how does the sensing distance depends on the object material.

14.4.3 Instruments used

- Inductive sensor MCPIP-T30L-011

- Capacitive sensor EC3025 PPAPL-1

- Hall sensor MM18-70APS-ZUK

- Optical sensor S15-PA-2-C10-NK

14.4.4 Procedure description

1. Turn on the power supply, and set the voltage to 24 V. The motor starts to spin; the rotating table starts to move.

2. Place different objects on the rotating table. The LEDs on the sensors indicate detection; find out if the object is detected and at what distance.

14.5   Position—linear-displacement sensors

14.5.1   Introduction

For a precise measurement of linear distance, sensors such as linear incremental encoders or inductosyn can be used. Position sensors are relative (incremental) or absolute. An inductosyn is working in the same way as a resolver, but the sensor is linear. It is a relative sensor. The SPIRO sensor (SPIRO is a trademark) work similarly. The movable sensor head is sensing a ferromagnetic spiral inside a stainless steel tube. The setup is shown in figure 14.7.

14.5.2   Tasks

1. Measure and plot in charts the steady-state characteristics of all used sensors. Use the Induktosyn sensor as a reference.
   Used sensors:
   a) sensor SPIRO
   b) laser sensor O1DLF3KG—output from the display and current output

FIGURE 14.7: Linear axis; inductosyn is from the back of the axis, not visible in photograph

2. For positions 50 mm and 100 mm from the initial position, show and record the histogram and standard deviation of the current output from the laser sensor.

14.5.3 Instruments used

- Linear axis with control electronics

- Induktosyn FERSYN 01, inductive, linear, cyclic absolute sensors with period 2 mm, resolution 1 µm, electronic unit BC 50 LM

- Limat SPIRO (the company now names the sensor ENDURRO), resolution 1 µm, electronic unit AXICA M

- Laser distance meter O1DLF3KG

- Power supplies

- Multimeter Keysight 34461 A

14.5.4 Procedure description

1. Change the position of the linear axis by means of the position controller. Measure the output of all sensors

2. For a constant position, display the histogram of the current output for the laser distance meter O1DLF3KG. The shown curve should be Gaussian.

References

[1] Poids et mesures, Bureau International des. *Evaluation of measurement data—Guide to the expression of uncertainty in measurement, GUM 1995 with minor corrections* [online] [visited on 2020-06-01]. Available from: `https://www.bipm.org/utils/common/documents/jcgm/JCGM_100_2008_E.pdf`.

[2] Hogan, R. *Probability Distributions for Measurement Uncertainty* [online]. 2015 [visited on 2020-06-01]. Available from: `https://www.isobudgets.com/probability-distributions-for-measurement-uncertainty/`.

[3] EPCOS. *Temperature Measurement B57164 Leaded Disks K164* [online] [visited on 2020-06-01]. Available from: `https://www.bucek.name/pdf/k164nk.pdf`.

[4] STMicroelectronics. *Precision very low power CMOS dual operational amplifiers* [online] [visited on 2020-06-01]. Available from: `https://www.st.com/content/ccc/resource/technical/document/datasheet/38/5b/ca/bd/c0/26/4a/34/CD00000880.pdf/files/CD00000880.pdf/jcr:content/translations/en.CD00000880.pdf`.

[5] Pro'skit. *MT-1232 manual* [online] [visited on 2020-06-01]. Available from: `https://www.manualsearcher.com/pros-kit/mt-1232/manual`.

[6] Jencik, J., Volf, J. a kol. *Technická měření*. Vydavatelství ČVUT, Praha, 2000 (in Czech).

[7] Preston-Thomas, H. *The International Temperature Scale of 1990 (ITS-90)* [online]. 1990 [visited on 2020-06-01]. Available from: `https://www.nist.gov/system/files/documents/pml/div685/grp01/ITS-90_metrologia.pdf`.

[8] *DIN IEC 751 Temperature/Resistance Table for Platinum Sensors* [online] [visited on 2020-06-01]. Available from: `https://www.lakeshore.com/docs/default-source/product-downloads/f038-00-00.pdf?sfvrsn=f24186e9_3`.

[9] *100 Ω Platinum RTD 0.003 92 coefficient temperature in C* [online] [visited on 2020-06-01]. Available from: `https://www.pyromation.com/Downloads/Data/392_c.pdf`.

[10] *RTD Nickel elements* [online]. 2019 [visited on 2020-06-01]. Available from: https://sensor-technology.com.br/index.php/elementos-rtd-de-niquel/?lang=en.

[11] *ITS90 Thermocouple tables* [online]. 2017 [visited on 2020-06-01]. Available from: https://srdata.nist.gov/its90/main/its90_main_page.html.

[12] *Revised Thermocouple Reference Tables, Type J* [online] [visited on 2020-06-01]. Available from: https://assets.omega.com/pdf/tables_and_graphs/thermocouple-type-j-celsius.pdf.

[13] *Revised Thermocouple Reference Tables, Type K* [online] [visited on 2020-06-01]. Available from: https://assets.omega.com/pdf/tables_and_graphs/thermocouple-type-k-celsius.pdf.

[14] *Revised Thermocouple Reference Tables, Type T* [online] [visited on 2020-06-01]. Available from: https://assets.omega.com/pdf/tables_and_graphs/thermocouple-type-t-celsius.pdf.

[15] *Revised Thermocouple Reference Tables, Type E* [online] [visited on 2020-06-01]. Available from: https://assets.omega.com/pdf/tables_and_graphs/thermocouple-type-e-celsius.pdf.

[16] *Revised Thermocouple Reference Tables, Type N* [online] [visited on 2020-06-01]. Available from: https://assets.omega.com/pdf/tables_and_graphs/thermocouple-type-n-celsius.pdf.

[17] *Revised Thermocouple Reference Tables, Type R* [online] [visited on 2020-06-01]. Available from: https://assets.omega.com/pdf/tables_and_graphs/thermocouple-type-r-celsius.pdf.

[18] *Revised Thermocouple Reference Tables, Type S* [online] [visited on 2020-06-01]. Available from: https://assets.omega.com/pdf/tables_and_graphs/thermocouple-type-s-celsius.pdf.

[19] *Wire Color Codes and Limits of Error* [online] [visited on 2020-06-01]. Available from: https://www.omega.co.uk/techref/colorcodes.html.

[20] *Revised Thermocouple Reference Tables, Type B* [online] [visited on 2020-06-01]. Available from: https://assets.omega.com/pdf/tables_and_graphs/thermocouple-type-b-celsius.pdf.

[21] *Temperature Sensor Installation for Best Response and Accuracy* [online]. 2017 [visited on 2020-06-01]. Available from: https://blog.isa.org/temperature-sensor-installation-best-response-accuracy.

[22] Ripka, P., Tipek, A., Vinci Project, Leonardo da. *Master Book on Sensors: Modular Courses on Modern Sensors—Leonardo Da Vinci Project CZ/PP-134026*. BEN—technical literature for Skoda Auto, 2003. ISBN 9788073001292. Available also from: https://books.google.cz/books?id=dzgmAAAACAAJ.

[23] Chandos, R.J., Chandos, R.E. *Radiometric Properties of Isothermal Diffuse Wall Cavity Sources* [online]. 2013 [visited on 2020-06-01]. Available from: https://www.osapublishing.org/ao/abstract.cfm?uri=ao-13-9-2142.

[24] Zhang, Y. Effects of Temperature on the Spectral Emissivity of C/SiC Composites. *Ceramics-Silikáty*. 2016, vol. 60, no. 2, pp. 152–155.

[25] Bohren, R., C. F. Huffman D. *Absorption and Scattering of Light by Small Particles*. Wiley, 2007.

[26] Vavricka, R. *Informační příručka pro projektanty—kondenzační kotle, kontrla kotrů a příprava teplé vody (in Czech)*. 2018.

[27] Lord, S.D. *NASA Technical Memorandum 103957*, Technical Memorandum. NASA. [online]. 1992 [visited on 2020-06-01]. Available from: https://www.gemini.edu/observing/telescopes-and-sites/sites#Transmission.

[28] *Absorption by Atmospheric Gases of Incoming and Outgoing Radiation* [online] [visited on 2020-06-01]. Available from: https://cleanet.org/clean/community/activities/c4.html.

[29] *Common Infrared Optical Materials and Coatings: A Guide to Properties, Performance and Applications* [online]. 2017 [visited on 2020-06-01]. Available from: https://www.photonics.com/a25495/Common_Infrared_Optical_Materials_and_Coatings_A.

[30] *Optical Materials* [online]. 2017 [visited on 2020-06-01]. Available from: https://www.ii-vi.com/product-category/products/optics/optical-materials/.

[31] *Materials for Infrared Optics* [online]. 2017 [visited on 2020-06-01]. Available from: https://wp.optics.arizona.edu/optomech/wp-content/uploads/sites/53/2016/10/Saayman-521-Tutorial.pdf.

[32] *RADIATION PYROMETERS* [online]. 2017 [visited on 2020-06-01]. Available from: https://optron.de/measuring-technology/pyrometers/?lang=en.

[33] *Stationary pyrometers Two-colour pyrometer PX 40* [online]. 2020 [visited on 2020-06-01]. Available from: https://www.keller.de/en/its/pyrometers/stationary-pyrometers/cellatemp-px/cellatemp-px-40.htm.

[34] Hoffman, K. *An Introduction to Measurement using Strain Gages*. Hottinger Baldwin Messtechnik GmbH, Darmstadt, Germany, 1989.

[35] *Sensitivity of Strain Gage Wire Materials* [online]. 2017 [visited on 2020-06-01]. Available from: https://www.efunda.com/designstandards/sensors/strain_gages/strain_gage_sensitivity.cfm.

[36] *Manufacture of strain gauges & transducers VTS Zlín* [online]. 2017 [visited on 2020-06-01]. Available from: https://vtsz.cz/image.ashx?i=165315.pdf&fn=.

[37]   *Residual Stress Determination* [online]. 2017 [visited on 2020-06-01]. Available from: http://www.stresscraft.co.uk/residual_stress_ determination.htm.

[38]   *Rotierende Drehmomentsensoren mit berührungsloser Übertragung* [online]. 2017 [visited on 2020-06-01]. Available from: https : / / www . lorenz – messtechnik . de / deutsch / produkte / drehmoment _ rotierend_schleifringlos.php.

[39]   *Permittivity* [online] [visited on 2020-06-01]. Available from: http:// maxwells-equations.com/materials/permittivity.php.

[40]   Dennis, J.O., Ahmad, F., Khir, M.H. CMOS Compatible Bulk Micromachining. In: Takahata, K. (ed.). *Advances in Micro/Nano Electromechanical Systems and Fabrication Technologies.* Rijeka: InTech, 2013, chap. 05. Available from DOI: 10.5772/55526.

[41]   *Optical Encoder Wheel Generator* [online] [visited on 2020-06-01]. Available from: http : / / www . bushytails . net / ~randyg / encoder / encoderwheel.html.

[42]   Etinger, A., Litvak, B., Pinhasi, Y. Multi Ray Model for Near-Ground Millimeter Wave Radar. *Sensors* [online]. 2017, vol. 17 [visited on 2020-06-01]. Available from DOI: 10.3390/s17091983.

[43]   *Operating Principles for Inductive Proximity Sensors* [online]. 2017 [visited on 2020-06-01]. Available from: http://www.fargocontrols.com/ sensors/inductive_op.html.

[44]   *Operating Principles for Capacitive Proximity Sensors* [online]. 2017 [visited on 2020-06-01]. Available from: http://www.fargocontrols. com/sensors/capacitive_op.html.

[45]   M., Novák. *New Methods for Instantaneous Angular Velocity Measurement by Analog Signal Processing.* 2008. Ph.D. Thesis. Czech Technical University in Prague, Faculty of Mechanical Engineering.

[46]   Mohd, H., Khir, P.Q., Qu, H. A Low-Cost CMOS-MEMS Piezoresistive Accelerometer with Large Proof Mass. *Sensors* [online]. 2011, vol. 11 [visited on 2018-01-10]. Available from DOI: 10.3390/s110807892.

[47]   Qu, P. and Qu, H. Design and Characterization of a Fully Differential MEMS Accelerometer Fabricated Using MetalMUMPs Technology. *Sensors* [online]. 2013, vol. 13 [visited on 2018-01-10]. Available from DOI: 10.3390/s130505720.

[48]   Zhou, Wu. Material Viscoelasticity-Induced Drift of Micro-Accelerometers. *Materials* [online]. 2017, vol. 10 [visited on 2018-01-10]. Available from DOI: 10.3390/ma10091077.

[49]   Kalenik, J. and Pajak, R. A cantilever optical-fiber accelerometer. *Sensors and Actuators A.* 1998, vol. 68.

[50] *Temperature Measurement During the Rod Heat Treatment* [online]. 2017 [visited on 2020-06-01]. Available from: https://www.instrumart.com/products/40221/dh-budenberg-cpb3800-hydraulic-deadweight-tester.

[51] *DOS001* [online]. 2017 [visited on 2020-06-01]. Available from: https://stiko.com/products/calibration-equipment/deadweight-testers/.

[52] *McLeod gauge* [online] [visited on 2020-06-01]. Available from: https://commons.wikimedia.org/wiki/File:McLeod_gauge_01.jpg.

[53] Ni, Z. Monolithic Composite Pressure + Acceleration + Temperature + Infrared Sensor Using a Versatile Single-Sided SiN/Poly-Si/Al Process-Module. *Sensors* [online]. 2012, vol. 13 [visited on 2018-01-12]. Available from DOI: 10.3390/s130101085.

[54] Zhang, J. et al. Design Optimization and Fabrication of High-Sensitivity SOI Pressure Sensors with High Signal-to-Noise Ratios Based on Silicon Nanowire Piezoresistors. *Micromachines* [online]. 2016, vol. 7 [visited on 2018-01-12]. Available from DOI: 10.3390/mi7100187.

[55] Lung-Tai, C., Jin-Sheng, C., Chung-Yi, H. and Cheng, W.-H. Fabrication and Performance of MEMS-Based Pressure Sensor Packages Using Patterned Ultra-Thick Photoresists. *Sensors* [online]. 2009, vol. 9 [visited on 2018-01-12]. Available from DOI: 10.3390/s90806200.

[56] Cao, G., Xiaoping, W., Yong, X., Liu, S. A Micromachined Piezoresistive Pressure Sensor with a Shield Layer. *Sensors* [online]. 2016, vol. 16 [visited on 2018-01-12]. Available from DOI: 10.3390/s16081286.

[57] Li, J. A Temperature Compensation Method for Piezo-Resistive Pressure Sensor Utilizing Chaotic Ions Motion Algorithm Optimized Hybrid Kernel LSSVM. *Sensors* [online]. 2016, vol. 16 [visited on 2018-01-12]. Available from DOI: 10.3390/s16101707.

[58] Dai, C.L., Tai, Y.W., Kao, P.H. Modeling and Fabrication of Micro FET Pressure Sensor with Circuits. *Sensors* [online]. 2007, vol. 7(12) [visited on 2018-01-12]. Available from DOI: 10.3390/s7123386.

[59] *DH-Budenberg CPB3800 Hydraulic Deadweight Tester* [online]. 2017 [visited on 2020-06-01]. Available from: https://arunmicro.com/products/?product=pirani-gauge-heads-and-leads.

[60] *KJLC 345 Series Pirani Gauges 945 Controller* [online]. 2017 [visited on 2020-06-01]. Available from: https://www.lesker.com/newweb/gauges/pirani_kjlc_945.cfm.

[61] *Teorie vlhkého vzduchu (I)* [online]. 2017 (in Czech) [visited on 2020-06-01]. Available from: https://vetrani.tzb-info.cz/teorie-a-vypocty-vetrani-klimatizace/3323-teorie-vlhkeho-vzduchu-i.

[62] Lawrence, M.G. The relationship between relative humidity and the dewpoint temperature in moist air: A simple conversion and applications. *Bull. Amer. Meteor. Soc.* 2005, vol. 86, pp. 225–233.

[63] *Installation & Maintenance Instructions Optica* [online]. 2012 [visited on 2020-06-01]. Available from: `https://able.co.uk/media/2014/05/optica-iom.pdf`.

[64] *DPT-2011 Single Stage Dew Point Transmitter* [online]. 2017 [visited on 2020-06-01]. Available from: `http://www.yesinc.com/index.php/8-products/73-dpt-2011-html`.

[65] Jachowicz, R.S., Weremczuk, J., Paczesny, D., and Tarapata, G. A MEMS-based super fast dew point hygrometer?construction and medical applications. *Measurement Science and Technology.* 2009, vol. 20, no. 12. Available also from: `https://iopscience.iop.org/article/10.1088/0957-0233/20/12/124008`.

[66] Ogawa, A. *PsychrometricChart.SeaLevel.SI* [online]. 2012 [visited on 2020-06-01]. Available from: `https://commons.wikimedia.org/wiki/File:PsychrometricChart.SeaLevel.SI.svg`.

[67] *Digital Humidity Sensor SHT2x (RH/T)* [online]. 2017 [visited on 2020-06-01]. Available from: `https : / / www . sensirion . com / en / environmental - sensors / humidity - sensors / humidity - temperature-sensor-sht2x-digital-i2c-accurate/`.

[68] Tripathy, A. Role of Morphological Structure, Doping, and Coating of Different Materials in the Sensing Characteristics of Humidity Sensors. *Sensors* [online]. 2014, vol. 14(9) [visited on 2018-01-12]. Available from DOI: `10.3390/s140916343`.

[69] Liu, M.Q., Wang, C., Kim, N.Y. High-Sensitivity and Low-Hysteresis Porous MIMType Capacitive Humidity Sensor Using Functional Polymer Mixed with TiO2 Microparticles. *Sensors* [online]. 2017, vol. 17(2) [visited on 2018-01-12]. Available from DOI: `10.3390/s17020284`.

[70] Kumar, L., Islam, T., Mukhopadhyay, S.C. Sensitivity Enhancement of a PPM Level Capacitive Moisture Sensor. *Electronics* [online]. 2017, vol. 6(2) [visited on 2018-01-12]. Available from DOI: `10.3390/electronics6020041`.

[71] *Orifice Plate Flowmeters for Steam, Liquids and Gases* [online]. 2017 [visited on 2020-06-01]. Available from: `https : / / assets . website-files.com/5ac7cf1999758e25761dac7f/5afb51fdc94885b63759f973_Spirax%20Sarco%20oriface%20plate.pdf`.

[72] *OPTISONIC 6300 P* [online]. 2020 [visited on 2020-06-01]. Available from: `https : / / krohne . com / en / products / flow - measurement / flowmeters/ultrasonic-flowmeters/optisonic-6300-p/`.

[73] *Mass Flow Meter* [online]. 2017 [visited on 2018-01-14]. Available from: `https://en.wikipedia.org/wiki/Mass_flow_meter`.

[74] *Fluke 576 Precision Infrared Thermometer—Users Manual* [online]. 2020 [visited on 2020-06-01]. Available from: `https://www.manualowl.com/m/Fluke/576/Manual/340141`.

[75] *Technical Information Data Bulletin, Pt 100* [online]. 2020 [visited on 2020-06-01]. Available from: `https://www.tnp-instruments.com/sitebuildercontent/sitebuilderfiles/pt100_385c_table.pdf`.

[76] *Thermometricscorp, 100 ohm Platinum RTD 0.00392 coefficient* [online]. 2020 [visited on 2020-06-01]. Available from: `https://www.thermometricscorp.com/PDFs/100-ohm-plt-rtd-0.00392-in-C.PDF`.

[77] *Resistance Value Chart per 1 deg C PT500* [online]. 2020 [visited on 2020-06-01]. Available from: `https://hayashidenko.co.jp/en/info13.html`.

[78] *Resistance Value Chart per 1 deg C PT1000* [online]. 2020 [visited on 2020-06-01]. Available from: `https://hayashidenko.co.jp/en/info14.html`.

[79] *120 ohm Nickel RTD-0,00682 coefficient* [online] [visited on 2020-06-01]. Available from: `https://www.thermometricscorp.com/PDFs/120_ohm_nickel_rtd-0.00672_in_C.PDF`.

[80] *120 Ω Nickel RTD 0.006 72 coefficient* [online]. 2020 [visited on 2020-06-01]. Available from: `https://www.pyromation.com/Downloads/Data/672_c.pdf`.

[81] *10 ohm Copper RTD - 0,00427 coefficient* [online] [visited on 2020-06-01]. Available from: `http://www.cromptonusa.com/10RTD_C.pdf`.

[82] *10Ω Copper RTD 0.00427 coefficient* [online]. 2020 [visited on 2020-06-01]. Available from: `http://www.cromptonusa.com/10RTD_C.pdf`.

[83] *Tables of Thermoelectric Voltages and Coefficients for Download* [online]. 2017 [visited on 2020-06-01]. Available from: `https://srdata.nist.gov/its90/download/download.html`.

[84] *Emissivity Table* [online]. 2019 [visited on 2020-06-01]. Available from: `http://www.thermocera.com/_src/2952/emissivity20tables.pdf`.

[85] *Table of emissivity of various surfaces* [online]. 2019 [visited on 2020-06-01]. Available from: `http://www-eng.lbl.gov/~dw/projects/DW4229_LHC_detector_analysis/calculations/emissivity2.pdf`.

[86] *Emissivity Coefficients Materials* [online]. 2019 [visited on 2020-06-01]. Available from: `https://www.engineeringtoolbox.com/emissivity-coefficients-d_447.html`.

[87] *Emissivity Values of Common Materials* [online]. 2019 [visited on 2020-06-01]. Available from: `https://www.bergeng.com/mm5/downloads/fluke/Emissivity-Values-of-Common-Materials-Chart.pdf`.

Other recommended literature

- *Orifice plate flowmeters for steam, liquids and gases* [online]. 2017 [cit. 2017-12-27]. Available from: `https://imistorage.blob.core.windows.net/imidocs/0699p031%20orifice%20plate%20flowmeters.pdf`.

- S., Bell. *A Beginner's Guide to Uncertainty of Measurement* [online] [cit. 2017-12-26]. Available from: `https://www.npl.co.uk/special-pages/guides/gpg118_begguide2measure`.

- *Wire Color Codes and Limits of Error* [online]. 1990 [cit. 2017-12-18]. Available from: `https://www.omega.co.uk/techref/colorcodes.html`.

- *Cold Junction Compensator* [online] [cit. 2017-12-18]. Available from: `https://sea.omega.com/th/pptst/CJ_MODULE.html`.

- *Fundamentals of Building a Test System* [online] [cit. 2020-01-19]. Available from: `https://www.ni.com/cs-cz/shop/pxi/fundamentals-of-building-a-test-system.html`.

- C., Dunn W. *Fundamentals of Industrial Instrumentation and Process Control*. McGraw-Hill, 2005.

- S., Morris Alan. *Measurement and Instrumentation Principles*. Butterworth-Heinemann, 2001.

- G., Webster J. *The Measurement, Instrumentation and Sensors Handbook*. CRC Press, 1999.

- B., Lipták. *Process Measurement and Analysis*. CRC Press, 2003.

- A., Nebylov. *Aerospace Sensors, Sensor technology series*. Momentum Press, 2012.

- *MAX4194 MAX4197 Micropower, Single-Supply, Rail-to-Rail, Precision Instrumentation Amplifiers* [online]. 2017 [cit. 2017-12-18]. Available from: `https://datasheets.maximintegrated.com/en/ds/MAX4194-MAX4197.pdf`.

- (EDITOR), Jung W. *Op Amp Applications Handbook* [online]. 2017 [cit. 2020-01-19]. Available from: `https://www.analog.com/en/education/education-library/op-amp-applications-handbook.html#`.

- A., Huges T. *Measurement and Control Basics, 3rd Edition*. ISA Press, 2002.

- S., Wilson Jon. *Sensor Technology Handbook*. Elsevier, 2005.

- J., Turner. *Automotive sensors*. Momentum Press, 2009.

- J., Szabo T. *Diagnostic Ultrasound Imaging: Inside Out 2nd Edition.* Academic Press, 2013.

- N., Cheeke J.David. *Fundamentals and Applications of Ultrasonic Waves.* CRC Press, 0202.

- B., Lipták. *Instrument Engineers' Handbook, Vol. 1: Process Measurement and Analysis 4th Edition.* CRC Press, 2003.

- *DEE-A Alloy Steel S Type Load Cell 50kg 20T* [online]. 2017 [cit. 2017-12-18]. Available from: http://www.coventryscale.co.uk/scale-type/load-cells/tension-s-typeload-cells/s-type-load-cell/.

- *Beam Load Cell* [online]. 2017 [cit. 2017-12-18]. Available from: https://www.omega.com/pptst/LC501.html.

- *Electrical Temperature Measurement with thermocouples and resistance thermometers, 2nd edition.*CRC Press, 2014.

- K., McMillan G. *Advanced Temperature Measurement and Control.* International Society of Automation,2011.

- *Practical Instrumentation for Automation and Process Control for Engineers and Technicians.*IDC Technologies, 2004.

- KIM, Youngdeuk. Capacitive humidity sensor design based on anodic aluminum oxide. *Sensors and Actuators B: Chemical.* 2009, roč. 141, č. 2, s. 441–446.

- SHAMALA, K.S. Characterization of Al2O3 thin films prepared by spray pyrolysis method for humidity sensor. *Sensors and Actuators A: Physical.* 2007, roč. 135, č. 2, s. 552–557.

- CHEN, Zhi; JIN, Mao-Chang. An alpha-alumina moisture sensor for relative and absolute humidity measurement. In: IEEE (ed.). *Conference Record of the 1992 IEEE Industry Applications Society Annual Meeting.* 1992, s. 1668–1675. Available from DOI: 11109/IAS.1992.244236.

- A., Arnau. *Piezoelectric Transducers and Applications - Second Edition.* Springer, 2008.

- *Introduction to Pt100 RTD Temperature Sensors*[online] [cit. 2020-01-19]. Available from: https://www.omega.com/en-us/resources/rtd.

- *An Explanation of the Beta and Steinhart-Hart Equations for Representing the Resistance vs. Temperature Relationship in NTC Thermistor Materials*[online] [cit. 2020-01-19]. Available from: https://www.qtisensing.com/wp-content/uploads/2019/09/Beta-vs-Steinhart-Hart.pdf.

- *Negative Temperature Coefficient Thermistors* [online] [cit. 2020-01-19]. Available from: https://www.qtisensing.com/wp-content/uploads/20 19/09/Whole-Article.pdf.

- *How to Measure the Unknown Thermal Emissivity of Objects/Materials Using the U5855A TrueIR Thermal Imager* [online] [cit. 2020-01-19]. Available from: https://www.keysight.com/zz/en/assets/7018-04614/app lication-notes/5992-0222.pdf.

- *Efficient Windows Collaborative - Window Technologies Low-E Coatings* [online] [cit. 2017-12-30]. Available from: https://c.ymcdn.com/sites/ www.rfmaonline.com/resource/resmgr/crfp/windowtechnologieslow-ecoati.pdf.

- *Principles of Non-Contact Temperature Measurement* [online] [cit. 2020-01-19]. Available from:http://support.fluke.com/raytek-sales/Dow nload/Asset/IR_THEORY_55514_ENG_REVB_LR.PDF.

- Fluke corporation. *Introduction to Thermography Principles.* Amer Technical Pub; 1 edition,2009.

- W., Ruddock R. *Basic Infrared Thermography Principles.* Reliability-web.com Press, 2011

- H., Kaplan. *Practical Applications of Infrared Thermal Sensing and Imaging Equipment, Third Edition (SPIE Tutorial Text Vol. TT75) (Tutorial Texts in Optical Engineering) 3rd Edition.* SPIE Publications, 2007.

- D., Lanzoni. *Infrared Thermography: electrical and industrial applications.* CreateSpace IndependentPublishing Platform, 2015.

- C., Holst G. *Common Sense Approach to Thermal Imaging.* SPIE-International Society for OpticalEngine, 2015.

- M., Volmer. *Infrared Thermal Imaging: Fundamentals, Research and Applications 1st Edition.*Wiley-VCH, 201

- *Thermal imaging guidebook for industrial applications* [online] [cit. 2020-01-19]. Available from: https://www.flir.com/discover/instruments/ thermal-imaging-guidebook-for-industrial-applications/.

- *Pirani gauges* [online] [cit. 2018-01-10]. Available from: http://thinfilm science.com/en-US/Pigauges.aspx.

- *G-TRAN Series Atmospheric Pirani Vaccum Gauge SW1* [online] [cit. 2018-01-12]. Available from: https://www.ulvac.co.jp/products_e/c omponents/vacuum-gauge/pirani-vacuum-gauge/sw1.

- Bello., I. *Vacuum and Ultravacuum: Physics and Technology.*CRC Press; 1 edition (November 9, 2017), ISBN: 978-1498782043.

- *How do I calculate dew point when I know the temperature and the relative humidity?* [online]. 2017 [cit. 2017-12-19]. Available from: `https://irid l.ldeo.columbia.edu/dochelp/QA/Basic/dewpoint.html`.

- *KJLC® 375 Series Panel Mount / Benchtop Gauge Controller* [online]. 2017 [cit. 2018-01-12]. Available from: `https://www.lesker.com/newwe b/gauges/convection_kjlc_5.cfm`.

- G., Korotcenkov. *Handbook of Humidity Measurement.*CRC Press; 1 edition (March 26, 2018), ISBN: 978-1138300217.

A

Pt100, DIN/EN/IEC 60751, $\alpha = 0.00385$

TABLE A.1: Pt100 $\alpha = 0.00385$, resistance in ohm vs. temperature in °C, negative temperatures, data [75]

°C	0	1	2	3	4	5	6	7	8	9
-200	18.520									
-190	22.826	22.397	21.967	21.538	21.108	20.677	20.247	19.815	19.384	18.952
-180	27.096	26.671	26.245	25.819	25.392	24.965	24.538	24.110	23.682	23.254
-170	31.335	30.913	30.490	30.067	29.643	29.220	28.796	28.371	27.947	27.522
-160	35.543	35.124	34.704	34.284	33.864	33.443	33.022	32.601	32.179	31.757
-150	39.723	39.306	38.889	38.472	38.055	37.637	37.219	36.800	36.382	35.963
-140	43.876	43.462	43.048	42.633	42.218	41.803	41.388	40.972	40.556	40.140
-130	48.005	47.593	47.181	46.769	46.356	45.944	45.531	45.118	44.704	44.290
-120	52.110	51.700	51.291	50.881	50.471	50.060	49.650	49.239	48.828	48.416
-110	56.193	55.786	55.378	54.970	54.562	54.154	53.746	53.337	52.928	52.519
-100	60.256	59.850	59.445	59.039	58.633	58.227	57.821	57.414	57.007	56.600
-90	64.300	63.896	63.492	63.088	62.684	62.280	61.876	61.471	61.066	60.661
-80	68.325	67.924	67.522	67.120	66.717	66.315	65.912	65.509	65.106	64.703
-70	72.335	71.934	71.534	71.134	70.733	70.332	69.931	69.530	69.129	68.727
-60	76.328	75.929	75.530	75.132	74.732	74.333	73.934	73.534	73.134	72.735
-50	80.306	79.909	79.512	79.114	78.717	78.319	77.921	77.523	77.125	76.726
-40	84.271	83.875	83.479	83.083	82.687	82.290	81.894	81.497	81.100	80.703
-30	88.222	87.827	87.433	87.038	86.643	86.248	85.853	85.457	85.062	84.666
-20	92.160	91.767	91.373	90.980	90.586	90.192	89.799	89.404	89.010	88.616
-10	96.086	95.694	95.302	94.909	94.517	94.124	93.732	93.339	92.946	92.553
0	100.000	99.609	99.218	98.827	98.436	98.044	97.653	97.261	96.870	96.478

TABLE A.2: Pt100 $\alpha = 0.00385$, resistance in ohm vs. temperature in °C, positive temperatures, data [75]

°C	0	1	2	3	4	5	6	7	8	9
0	100.000	100.391	100.781	101.172	101.562	101.953	102.343	102.733	103.123	103.513
10	103.903	104.292	104.682	105.071	105.460	105.849	106.238	106.627	107.016	107.405
20	107.794	108.182	108.570	108.959	109.347	109.735	110.123	110.510	110.898	111.286
30	111.673	112.060	112.447	112.835	113.221	113.608	113.995	114.382	114.768	115.155
40	115.541	115.927	116.313	116.699	117.085	117.470	117.856	118.241	118.627	119.012
50	119.397	119.782	120.167	120.552	120.936	121.321	121.705	122.090	122.474	122.858
60	123.242	123.626	124.009	124.393	124.777	125.160	125.543	125.926	126.309	126.692
70	127.075	127.458	127.840	128.223	128.605	128.987	129.370	129.752	130.133	130.515
80	130.897	131.278	131.660	132.041	132.422	132.803	133.184	133.565	133.946	134.326
90	134.707	135.087	135.468	135.848	136.228	136.608	136.987	137.367	137.747	138.126
100	138.505	138.885	139.264	139.643	140.022	140.400	140.779	141.158	141.536	141.914
110	142.293	142.671	143.049	143.426	143.804	144.182	144.559	144.937	145.314	145.691
120	146.068	146.445	146.822	147.198	147.575	147.951	148.328	148.704	149.080	149.456
130	149.832	150.208	150.583	150.959	151.334	151.710	152.085	152.460	152.835	153.210
140	153.584	153.959	154.333	154.708	155.082	155.456	155.830	156.204	156.578	156.952
150	157.325	157.699	158.072	158.445	158.818	159.191	159.564	159.937	160.309	160.682
160	161.054	161.427	161.799	162.171	162.543	162.915	163.286	163.658	164.030	164.401
170	164.772	165.143	165.514	165.885	166.256	166.627	166.997	167.368	167.738	168.108

Continues on next page

243

Pt100, DIN/EN/IEC 60751, α = 0.00385

TABLE A.2 – cont.

°C	0	1	2	3	4	5	6	7	8	9
180	168.478	168.848	169.218	169.588	169.958	170.327	170.696	171.066	171.435	171.804
190	172.173	172.542	172.910	173.279	173.648	174.016	174.384	174.752	175.120	175.488
200	175.856	176.224	176.591	176.959	177.326	177.693	178.060	178.427	178.794	179.161
210	179.528	179.894	180.260	180.627	180.993	181.359	181.725	182.091	182.456	182.822
220	183.188	183.553	183.918	184.283	184.648	185.013	185.378	185.743	186.107	186.472
230	186.836	187.200	187.564	187.928	188.292	188.656	189.019	189.383	189.746	190.110
240	190.473	190.836	191.199	191.562	191.924	192.287	192.649	193.012	193.374	193.736
250	194.098	194.460	194.822	195.183	195.545	195.906	196.268	196.629	196.990	197.351
260	197.712	198.073	198.433	198.794	199.154	199.514	199.875	200.235	200.595	200.954
270	201.314	201.674	202.033	202.393	202.752	203.111	203.470	203.829	204.188	204.546
280	204.905	205.263	205.622	205.980	206.338	206.696	207.054	207.411	207.769	208.127
290	208.484	208.841	209.198	209.555	209.912	210.269	210.626	210.982	211.339	211.695
300	212.052	212.408	212.764	213.120	213.475	213.831	214.187	214.542	214.897	215.252
310	215.608	215.962	216.317	216.672	217.027	217.381	217.736	218.090	218.444	218.798
320	219.152	219.506	219.860	220.213	220.567	220.920	221.273	221.626	221.979	222.332
330	222.685	223.038	223.390	223.743	224.095	224.447	224.799	225.151	225.503	225.855
340	226.206	226.558	226.909	227.260	227.612	227.963	228.314	228.664	229.015	229.366
350	229.716	230.066	230.417	230.767	231.117	231.467	231.816	232.166	232.516	232.865
360	233.214	233.564	233.913	234.262	234.610	234.959	235.308	235.656	236.005	236.353
370	236.701	237.049	237.397	237.745	238.093	238.440	238.788	239.135	239.482	239.829
380	240.176	240.523	240.870	241.217	241.563	241.910	242.256	242.602	242.948	243.294
390	243.640	243.986	244.331	244.677	245.022	245.367	245.713	246.058	246.403	246.747
400	247.092	247.437	247.781	248.125	248.470	248.814	249.158	249.502	249.845	250.189
410	250.533	250.876	251.219	251.562	251.906	252.248	252.591	252.934	253.277	253.619
420	253.962	254.304	254.646	254.988	255.330	255.672	256.013	256.355	256.696	257.038
430	257.379	257.720	258.061	258.402	258.743	259.083	259.424	259.764	260.105	260.445
440	260.785	261.125	261.465	261.804	262.144	262.483	262.823	263.162	263.501	263.840
450	264.179	264.518	264.857	265.195	265.534	265.872	266.210	266.548	266.886	267.224
460	267.562	267.900	268.237	268.574	268.912	269.249	269.586	269.923	270.260	270.597
470	270.933	271.270	271.606	271.942	272.278	272.614	272.950	273.286	273.622	273.957
480	274.293	274.628	274.963	275.298	275.633	275.968	276.303	276.638	276.972	277.307
490	277.641	277.975	278.309	278.643	278.977	279.311	279.644	279.978	280.311	280.644
500	280.978	281.311	281.643	281.976	282.309	282.641	282.974	283.306	283.638	283.971
510	284.303	284.634	284.966	285.298	285.629	285.961	286.292	286.623	286.954	287.285
520	287.616	287.947	288.277	288.608	288.938	289.268	289.599	289.929	290.258	290.588
530	290.918	291.247	291.577	291.906	292.235	292.565	292.894	293.222	293.551	293.880
540	294.208	294.537	294.865	295.193	295.521	295.849	296.177	296.505	296.832	297.160
550	297.487	297.814	298.142	298.469	298.795	299.122	299.449	299.775	300.102	300.428
560	300.754	301.080	301.406	301.732	302.058	302.384	302.709	303.035	303.360	303.685
570	304.010	304.335	304.660	304.985	305.309	305.634	305.958	306.282	306.606	306.930
580	307.254	307.578	307.902	308.225	308.549	308.872	309.195	309.518	309.841	310.164
590	310.487	310.810	311.132	311.454	311.777	312.099	312.421	312.743	313.065	313.386
600	313.708	314.029	314.351	314.672	314.993	315.314	315.635	315.956	316.277	316.597
610	316.918	317.238	317.558	317.878	318.198	318.518	318.838	319.157	319.477	319.796
620	320.115	320.435	320.754	321.073	321.391	321.710	322.029	322.347	322.666	322.984
630	323.302	323.620	323.938	324.256	324.573	324.891	325.208	325.526	325.843	326.160
640	326.477	326.794	327.110	327.427	327.744	328.060	328.376	328.692	329.008	329.324
650	329.640	329.956	330.271	330.587	330.902	331.217	331.533	331.848	332.162	332.477
660	332.792	333.106	333.421	333.735	334.049	334.363	334.677	334.991	335.305	335.619
670	335.932	336.246	336.559	336.872	337.185	337.498	337.811	338.123	338.436	338.748
680	339.061	339.373	339.685	339.997	340.309	340.621	340.932	341.244	341.555	341.867
690	342.178	342.489	342.800	343.111	343.422	343.732	344.043	344.353	344.663	344.973
700	345.284	345.593	345.903	346.213	346.522	346.832	347.141	347.451	347.760	348.069
710	348.378	348.686	348.995	349.303	349.612	349.920	350.228	350.536	350.844	351.152
720	351.460	351.768	352.075	352.382	352.690	352.997	353.304	353.611	353.918	354.224
730	354.531	354.837	355.144	355.450	355.756	356.062	356.368	356.674	356.979	357.285
740	357.590	357.896	358.201	358.506	358.811	359.116	359.420	359.725	360.029	360.334
750	360.638	360.942	361.246	361.550	361.854	362.158	362.461	362.765	363.068	363.371
760	363.674	363.977	364.280	364.583	364.886	365.188	365.491	365.793	366.095	366.397
770	366.699	367.001	367.303	367.604	367.906	368.207	368.508	368.810	369.111	369.412
780	369.712	370.013	370.314	370.614	370.914	371.215	371.515	371.815	372.115	372.414
790	372.714	373.013	373.313	373.612	373.911	374.210	374.509	374.808	375.107	375.406
800	375.704	376.002	376.301	376.599	376.897	377.195	377.493	377.790	378.088	378.385
810	378.683	378.980	379.277	379.574	379.871	380.167	380.464	380.761	381.057	381.353

Continues on next page

TABLE A.2 − cont.

°C	0	1	2	3	4	5	6	7	8	9
820	381.649	381.946	382.242	382.537	382.833	383.129	383.424	383.720	384.015	384.310
830	384.605	384.900	385.195	385.489	385.784	386.078	386.373	386.667	386.961	387.255
840	387.549	387.843	388.136	388.430	388.723	389.016	389.310	389.603	389.896	390.188
850	390.481									

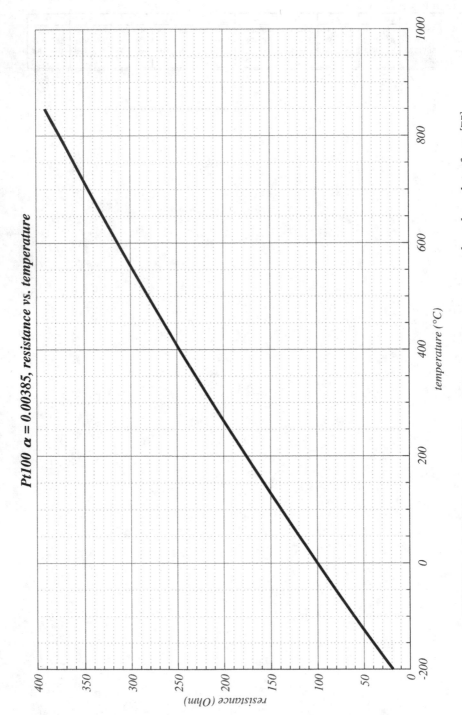

FIGURE A.1: Pt100 $\alpha = 0.00385$, resistance vs. temperature, based on data from [75]

B

Pt100, α = 0.00392

TABLE B.1: Pt100 $\alpha = 0.00392$, resistance in ohm vs. temperature in °C, negative temperatures, data [76]

°C	0	1	2	3	4	5	6	7	8	9
-200	17.08									
-190	21.46	21.02	20.58	20.15	19.71	19.27	18.83	18.40	17.96	17.52
-180	25.80	25.37	24.94	24.50	24.07	23.63	23.20	22.76	22.33	21.89
-170	30.11	29.68	29.25	28.82	28.39	27.96	27.53	27.10	26.67	26.23
-160	34.39	33.97	33.54	33.11	32.69	32.26	31.83	31.40	30.97	30.54
-150	38.65	38.22	37.80	37.37	36.95	36.52	36.10	35.67	35.25	34.82
-140	42.87	42.45	42.03	41.61	41.19	40.76	40.34	39.92	39.49	39.07
-130	47.07	46.66	46.24	45.82	45.40	44.98	44.56	44.14	43.72	43.29
-120	51.25	50.84	50.42	50.00	49.58	49.17	48.75	48.33	47.91	47.49
-110	55.41	54.99	54.58	54.16	53.75	53.33	52.92	52.50	52.09	51.67
-100	59.54	59.13	58.72	58.30	57.89	57.48	57.06	56.65	56.24	55.82
-90	63.66	63.25	62.84	62.43	62.01	61.60	61.19	60.78	60.37	59.96
-80	67.76	67.35	66.94	66.53	66.12	65.71	65.30	64.89	64.48	64.07
-70	71.84	71.43	71.02	70.61	70.21	69.80	69.39	68.98	68.57	68.17
-60	75.90	75.50	75.09	74.68	74.28	73.87	73.47	73.06	72.65	72.24
-50	79.95	79.55	79.14	78.74	78.33	77.93	77.52	77.12	76.71	76.31
-40	83.99	83.58	83.18	82.78	82.38	81.97	81.57	81.16	80.76	80.36
-30	88.01	87.61	87.21	86.80	86.40	86.00	85.60	85.20	84.79	84.39
-20	92.02	91.62	91.22	90.82	90.42	90.02	89.61	89.21	88.81	88.41
-10	96.02	95.62	95.22	94.82	94.42	94.02	93.62	93.22	92.82	92.42
0	100.00	99.60	99.20	98.81	98.41	98.01	97.61	97.21	96.81	96.41

TABLE B.2: Pt100 $\alpha = 0.00392$, resistance in ohm vs. temperature in °C, positive temperatures, data [76]

°C	0	1	2	3	4	5	6	7	8	9
0	100.00	100.40	100.80	101.19	101.59	101.99	102.39	102.78	103.18	103.58
10	103.97	104.37	104.77	105.16	105.56	105.95	106.35	106.75	107.14	107.54
20	107.93	108.33	108.72	109.12	109.52	109.91	110.30	110.70	111.09	111.49
30	111.88	112.28	112.67	113.07	113.46	113.85	114.25	114.64	115.03	115.43
40	115.82	116.21	116.61	117.00	117.39	117.79	118.18	118.57	118.96	119.35
50	119.75	120.14	120.53	120.92	121.31	121.71	122.10	122.49	122.88	123.27
60	123.66	124.05	124.44	124.83	125.22	125.61	126.00	126.39	126.78	127.17
70	127.56	127.95	128.34	128.73	129.12	129.51	129.90	130.29	130.68	131.07
80	131.45	131.84	132.23	132.62	133.01	133.39	133.78	134.17	134.56	134.95
90	135.33	135.72	136.11	136.49	136.88	137.27	137.65	138.04	138.43	138.81
100	139.20	139.59	139.97	140.36	140.74	141.13	141.51	141.90	142.29	142.67
110	143.06	143.44	143.83	144.21	144.59	144.98	145.36	145.75	146.13	146.52
120	146.90	147.28	147.67	148.05	148.43	148.82	149.20	149.58	149.97	150.35
130	150.73	151.11	151.50	151.88	152.26	152.64	153.02	153.41	153.79	154.17
140	154.55	154.93	155.31	155.70	156.08	156.46	156.84	157.22	157.60	157.98
150	158.36	158.74	159.12	159.50	159.88	160.26	160.64	161.02	161.40	161.78
160	162.16	162.54	162.91	163.29	163.67	164.05	164.43	164.81	165.19	165.56
170	165.94	166.32	166.70	167.07	167.45	167.83	168.21	168.58	168.96	169.34

Continues on next page

TABLE B.2 – cont.

°C	0	1	2	3	4	5	6	7	8	9
180	169.71	170.09	170.47	170.84	171.22	171.60	171.97	172.35	172.73	173.10
190	173.48	173.85	174.23	174.60	174.98	175.35	175.73	176.10	176.48	176.85
200	177.23	177.60	177.97	178.35	178.72	179.10	179.47	179.84	180.22	180.59
210	180.96	181.34	181.71	182.08	182.46	182.83	183.20	183.57	183.95	184.32
220	184.69	185.06	185.43	185.81	186.18	186.55	186.92	187.29	187.66	188.03
230	188.41	188.78	189.15	189.52	189.89	190.26	190.63	191.00	191.37	191.74
240	192.11	192.48	192.85	193.22	193.59	193.96	194.32	194.69	195.06	195.43
250	195.80	196.17	196.54	196.90	197.27	197.64	198.01	198.38	198.74	199.11
260	199.48	199.85	200.21	200.58	200.95	201.31	201.68	202.05	202.41	202.78
270	203.15	203.51	203.88	204.24	204.61	204.98	205.34	205.71	206.07	206.44
280	206.80	207.17	207.53	207.90	208.26	208.63	208.99	209.35	209.72	210.08
290	210.45	210.81	211.17	211.54	211.90	212.26	212.63	212.99	213.35	213.72
300	214.08	214.44	214.80	215.17	215.53	215.89	216.25	216.61	216.98	217.34
310	217.70	218.06	218.42	218.78	219.14	219.51	219.87	220.23	220.59	220.95
320	221.31	221.67	222.03	222.39	222.75	223.11	223.47	223.83	224.19	224.55
330	224.91	225.26	225.62	225.98	226.34	226.70	227.06	227.42	227.78	228.13
340	228.49	228.85	229.21	229.56	229.92	230.28	230.64	230.99	231.35	231.71
350	232.07	232.42	232.78	233.13	233.49	233.85	234.20	234.56	234.92	235.27
360	235.63	235.98	236.34	236.69	237.05	237.40	237.76	238.11	238.47	238.82
370	239.18	239.53	239.89	240.24	240.59	240.95	241.30	241.66	242.01	242.36
380	242.72	243.07	243.42	243.78	244.13	244.48	244.83	245.19	245.54	245.89
390	246.24	246.59	246.95	247.30	247.65	248.00	248.35	248.70	249.06	249.41
400	249.76	250.11	250.46	250.81	251.16	251.51	251.86	252.21	252.56	252.91
410	253.26	253.61	253.96	254.31	254.66	255.01	255.36	255.71	256.06	256.40
420	256.75	257.10	257.45	257.80	258.15	258.49	258.84	259.19	259.54	259.89
430	260.23	260.58	260.93	261.27	261.62	261.97	262.31	262.66	263.01	263.35
440	263.70	264.05	264.39	264.74	265.08	265.43	265.78	266.12	266.47	266.81
450	267.16	267.50	267.85	268.19	268.54	268.88	269.23	269.57	269.91	270.26
460	270.60	270.95	271.29	271.63	271.98	272.32	272.66	273.01	273.35	273.69
470	274.03	274.38	274.72	275.06	275.40	275.75	276.09	276.43	276.77	277.11
480	277.46	277.80	278.14	278.48	278.82	279.16	279.50	279.84	280.18	280.52
490	280.87	281.21	281.55	281.89	282.23	282.57	282.91	283.24	283.58	283.92
500	284.26	284.60	284.94	285.28	285.62	285.96	286.30	286.63	286.97	287.31
510	287.65	287.99	288.32	288.66	289.00	289.34	289.67	290.01	290.35	290.69
520	291.02	291.36	291.70	292.03	292.37	292.71	293.04	293.38	293.71	294.05
530	294.39	294.72	295.06	295.39	295.73	296.06	296.40	296.73	297.07	297.40
540	297.74	298.07	298.41	298.74	299.07	299.41	299.74	300.07	300.41	300.74
550	301.08	301.41	301.74	302.07	302.41	302.74	303.07	303.41	303.74	304.07
560	304.40	304.73	305.07	305.40	305.73	306.06	306.39	306.72	307.06	307.39
570	307.72	308.05	308.38	308.71	309.04	309.37	309.70	310.03	310.36	310.69
580	311.02	311.35	311.68	312.01	312.34	312.67	313.00	313.33	313.66	313.99
590	314.31	314.64	314.97	315.30	315.63	315.96	316.28	316.61	316.94	317.27
600	317.59	317.92	318.25	318.58	318.90	319.23	319.56	319.88	320.21	320.54
610	320.86	321.19	321.52	321.84	322.17	322.49	322.82	323.14	323.47	323.79
620	324.12	324.44	324.77	325.09	325.42	325.74	326.07	326.39	326.72	327.04
630	327.36	327.69	328.01	328.34	328.66	328.98	329.31	329.63	329.95	330.28
640	330.60	330.92	331.24	331.57	331.89	332.21	332.53	332.85	333.18	333.50
650	333.82	334.14	334.46	334.78	335.11	335.43	335.75	336.07	336.39	336.71
660	337.03									

C

Pt500

TABLE C.1: Pt500, resistance in ohm vs. temperature in °C, negative temperatures, data [77]

°C	0	1	2	3	4	5	6	7	8	9
-70	361.65	359.65	357.65	355.65	353.65	351.65	349.65	347.65	345.65	343.65
-60	381.65	379.65	377.65	375.65	373.65	371.65	369.65	367.65	365.65	363.65
-50	401.55	399.55	397.55	395.55	393.6	391.6	389.6	387.6	385.6	383.65
-40	421.35	419.35	417.4	415.4	413.45	411.45	409.45	407.5	405.5	403.5
-30	441.1	439.15	437.15	435.2	433.2	431.25	429.25	427.3	425.3	423.35
-20	460.8	458.85	456.85	454.9	452.95	450.95	449	447	445.05	443.1
-10	480.45	478.45	476.5	474.55	472.6	470.6	468.65	466.7	464.75	462.75
0	500	498.05	496.1	494.15	492.2	490.2	488.25	486.3	484.35	482.4

TABLE C.2: Pt500, resistance in ohm vs. temperature in °C, positive temperatures, data [77]

°C	0	1	2	3	4	5	6	7	8	9
0	500	501.95	503.9	505.85	507.8	509.75	511.7	513.65	515.6	517.55
10	519.5	521.45	523.4	525.35	527.3	529.25	531.2	533.15	535.1	537
20	538.95	540.9	542.85	544.8	546.75	548.65	550.6	552.55	554.5	556.45
30	558.35	560.3	562.25	564.15	566.1	568.05	570	571.9	573.85	575.75
40	577.7	579.65	581.55	583.5	585.4	587.35	589.3	591.2	593.15	595.05
50	597	598.9	600.85	602.75	604.7	606.6	608.55	610.45	612.35	614.3
60	616.2	618.15	620.05	621.95	623.9	625.8	627.7	629.65	631.55	633.45
70	635.4	637.3	639.2	641.1	643.05	644.95	646.85	648.75	650.65	652.6
80	654.5	656.4	658.3	660.2	662.1	664	665.9	667.85	669.75	671.65
90	673.55	675.45	677.35	679.25	681.15	683.05	684.95	686.85	688.75	690.65
100	692.55	694.4	696.3	698.2	700.1	702	703.9	705.8	707.7	709.55
110	711.45	713.35	715.25	717.15	719	720.9	722.8	724.7	726.55	728.45
120	730.35	732.2	734.1	736	737.85	739.75	741.65	743.5	745.4	747.3
130	749.15	751.05	752.9	754.8	756.65	758.55	760.4	762.3	764.15	766.05
140	767.9	769.8	771.65	773.55	775.4	777.3	779.15	781	782.9	784.75
150	786.65	788.5	790.35	792.25	794.1	795.95	797.8	799.7	801.55	803.4
160	805.25	807.15	809	810.85	812.7	814.55	816.45	818.3	820.15	822
170	823.85	825.7	827.55	829.45	831.3	833.15	835	836.85	838.7	840.55
180	842.4	844.25	846.1	847.95	849.8	851.65	853.5	855.35	857.15	859
190	860.85	862.7	864.55	866.4	868.25	870.1	871.9	873.75	875.6	877.45
200	879.3	881.1	882.95	884.8	886.65	888.45	890.3	892.15	893.95	895.8
210	897.65	899.45	901.3	903.15	904.95	906.8	908.6	910.45	912.3	914.1
220	915.95	917.75	919.6	921.4	923.25	925.05	926.9	928.7	930.55	932.35
230	934.2	936	937.8	939.65	941.45	943.3	945.1	946.9	948.75	950.55
240	952.35	954.2	956	957.8	959.6	961.45	963.25	965.05	966.85	968.7
250	970.5	972.3	974.1	975.9	977.75	979.55	981.35	983.15	984.95	986.75
260	988.55	990.35	992.15	993.95	995.75	997.55	999.35	1001.15	1002.95	1004.75
270	1006.55	1008.35	1010.15	1011.95	1013.75	1015.55	1017.35	1019.15	1020.95	1022.75
280	1024.5	1026.3	1028.1	1029.9	1031.7	1033.5	1035.25	1037.05	1038.85	1040.65
290	1042.4	1044.2	1046	1047.8	1049.55	1051.35	1053.15	1054.9	1056.7	1058.5
300	1060.25	1062.05	1063.8	1065.6	1067.4	1069.15	1070.95	1072.7	1074.5	1076.25

Continues on next page

TABLE C.2 – cont.

°C	0	1	2	3	4	5	6	7	8	9
310	1078.05	1079.8	1081.6	1083.35	1085.15	1086.9	1088.7	1090.45	1092.2	1094
320	1095.75	1097.55	1099.3	1101.05	1102.85	1104.6	1106.35	1108.15	1109.9	1111.65
330	1113.4	1115.2	1116.95	1118.7	1120.45	1122.25	1124	1125.75	1127.5	1129.25
340	1131.05	1132.8	1134.55	1136.3	1138.05	1139.8	1141.55	1143.3	1145.1	1146.85
350	1148.6	1150.35	1152.1	1153.85	1155.6	1157.35	1159.1	1160.85	1162.6	1164.35
360	1166.05	1167.8	1169.55	1171.3	1173.05	1174.8	1176.55	1178.3	1180	1181.75
370	1183.5	1185.25	1187	1188.7	1190.45	1192.2	1193.95	1195.65	1197.4	1199.15
380	1200.9	1202.6	1204.35	1206.1	1207.8	1209.55	1211.3	1213	1214.75	1216.45
390	1218.2	1219.95	1221.65	1223.4	1225.1	1226.85	1228.55	1230.3	1232	1233.75
400	1235.45	1237.2	1238.9	1240.65	1242.35	1244.05	1245.8	1247.5	1249.25	1250.95
410	1252.65	1254.4	1256.1	1257.8	1259.55	1261.25	1262.95	1264.65	1266.4	1268.1
420	1269.8	1271.5	1273.25	1274.95	1276.65	1278.35	1280.05	1281.75	1283.5	1285.2
430	1286.9	1288.6	1290.3	1292	1293.7	1295.4	1297.1	1298.8	1300.5	1302.2
440	1303.9	1305.6	1307.3	1309	1310.7	1312.4	1314.1	1315.8	1317.5	1319.2
450	1320.9	1322.6	1324.3	1326	1327.65	1329.35	1331.05	1332.75	1334.45	1336.1
460	1337.8	1339.5	1341.2	1342.85	1344.55	1346.25	1347.95	1349.6	1351.3	1353
470	1354.65	1356.35	1358.05	1359.7	1361.4	1363.05	1364.75	1366.45	1368.1	1369.8
480	1371.45	1373.15	1374.8	1376.5	1378.15	1379.85	1381.5	1383.2	1384.85	1386.55
490	1388.2	1389.9	1391.55	1393.2	1394.9	1396.55	1398.2	1399.9	1401.55	1403.2
500	1404.9	1406.55	1408.2	1409.9	1411.55	1413.2	1414.85	1416.55	1418.2	1419.85

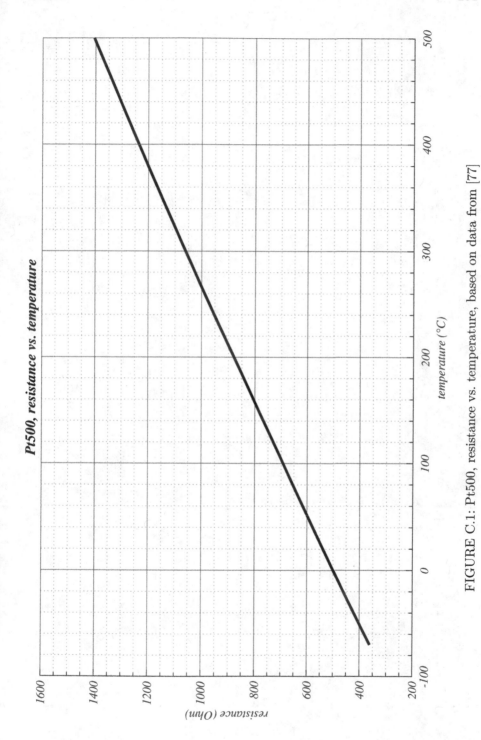

FIGURE C.1: Pt500, resistance vs. temperature, based on data from [77]

D

Pt1000

TABLE D.1: Pt1000, resistance in ohm vs. temperature in °C, negative
temperatures, data [78]

°C	0	1	2	3	4	5	6	7	8	9
-70	723.3	719.3	715.3	711.3	707.3	703.3	699.3	695.3	691.3	687.3
-60	763.3	759.3	755.3	751.3	747.3	743.3	739.3	735.3	731.3	727.3
-50	803.1	799.1	795.1	791.1	787.2	783.2	779.2	775.2	771.2	767.3
-40	842.7	838.7	834.8	830.8	826.9	822.9	818.9	815	811	807
-30	882.2	878.3	874.3	870.4	866.4	862.5	858.5	854.6	850.6	846.7
-20	921.6	917.7	913.7	909.8	905.9	901.9	898	894	890.1	886.2
-10	960.9	956.9	953	949.1	945.2	941.2	937.3	933.4	929.5	925.5
0	1000	996.1	992.2	988.3	984.4	980.4	976.5	972.6	968.7	964.8

TABLE D.2: Pt1000, resistance in ohm vs. temperature in °C, positive
temperatures, data [78]

°C	0	1	2	3	4	5	6	7	8	9
0	1000	1003.9	1007.8	1011.7	1015.6	1019.5	1023.4	1027.3	1031.2	1035.1
10	1039	1042.9	1046.8	1050.7	1054.6	1058.5	1062.4	1066.3	1070.2	1074
20	1077.9	1081.8	1085.7	1089.6	1093.5	1097.3	1101.2	1105.1	1109	1112.9
30	1116.7	1120.6	1124.5	1128.3	1132.2	1136.1	1140	1143.8	1147.7	1151.5
40	1155.4	1159.3	1163.1	1167	1170.8	1174.7	1178.6	1182.4	1186.3	1190.1
50	1194	1197.8	1201.7	1205.5	1209.4	1213.2	1217.1	1220.9	1224.7	1228.6
60	1232.4	1236.3	1240.1	1243.9	1247.8	1251.6	1255.4	1259.3	1263.1	1266.9
70	1270.8	1274.6	1278.4	1282.2	1286.1	1289.9	1293.7	1297.5	1301.3	1305.2
80	1309	1312.8	1316.6	1320.4	1324.2	1328	1331.8	1335.7	1339.5	1343.3
90	1347.1	1350.9	1354.7	1358.5	1362.3	1366.1	1369.9	1373.7	1377.5	1381.3
100	1385.1	1388.8	1392.6	1396.4	1400.2	1404	1407.8	1411.6	1415.4	1419.1
110	1422.9	1426.7	1430.5	1434.3	1438	1441.8	1445.6	1449.4	1453.1	1456.9
120	1460.7	1464.4	1468.2	1472	1475.7	1479.5	1483.3	1487	1490.8	1494.6
130	1498.3	1502.1	1505.8	1509.6	1513.3	1517.1	1520.8	1524.6	1528.3	1532.1
140	1535.8	1539.6	1543.3	1547.1	1550.8	1554.6	1558.3	1562	1565.8	1569.5
150	1573.3	1577	1580.7	1584.5	1588.2	1591.9	1595.6	1599.4	1603.1	1606.8
160	1610.5	1614.3	1618	1621.7	1625.4	1629.1	1632.9	1636.6	1640.3	1644
170	1647.7	1651.4	1655.1	1658.9	1662.6	1666.3	1670	1673.7	1677.4	1681.1
180	1684.8	1688.5	1692.2	1695.9	1699.6	1703.3	1707	1710.7	1714.3	1718
190	1721.7	1725.4	1729.1	1732.8	1736.5	1740.2	1743.8	1747.5	1751.2	1754.9
200	1758.6	1762.2	1765.9	1769.6	1773.3	1776.9	1780.6	1784.3	1787.9	1791.6
210	1795.3	1798.9	1802.6	1806.3	1809.9	1813.6	1817.2	1820.9	1824.6	1828.2
220	1831.9	1835.5	1839.2	1842.8	1846.5	1850.1	1853.8	1857.4	1861.1	1864.7
230	1868.4	1872	1875.6	1879.3	1882.9	1886.6	1890.2	1893.8	1897.5	1901.1
240	1904.7	1908.4	1912	1915.6	1919.2	1922.9	1926.5	1930.1	1933.7	1937.4
250	1941	1944.6	1948.2	1951.8	1955.5	1959.1	1962.7	1966.3	1969.9	1973.5
260	1977.1	1980.7	1984.3	1987.9	1991.5	1995.1	1998.7	2002.3	2005.9	2009.5
270	2013.1	2016.7	2020.3	2023.9	2027.5	2031.1	2034.7	2038.3	2041.9	2045.5
280	2049	2052.6	2056.2	2059.8	2063.4	2067	2070.5	2074.1	2077.7	2081.3
290	2084.8	2088.4	2092	2095.6	2099.1	2102.7	2106.3	2109.8	2113.4	2117
300	2120.5	2124.1	2127.6	2131.2	2134.8	2138.3	2141.9	2145.4	2149	2152.5

Continues on next page

Pt1000

TABLE D.2 – cont.

°C	0	1	2	3	4	5	6	7	8	9
310	2156.1	2159.6	2163.2	2166.7	2170.3	2173.8	2177.4	2180.9	2184.4	2188
320	2191.5	2195.1	2198.6	2202.1	2205.7	2209.2	2212.7	2216.3	2219.8	2223.3
330	2226.8	2230.4	2233.9	2237.4	2240.9	2244.5	2248	2251.5	2255	2258.5
340	2262.1	2265.6	2269.1	2272.6	2276.1	2279.6	2283.1	2286.6	2290.2	2293.7
350	2297.2	2300.7	2304.2	2307.7	2311.2	2314.7	2318.2	2321.7	2325.2	2328.7
360	2332.1	2335.6	2339.1	2342.6	2346.1	2349.6	2353.1	2356.6	2360	2363.5
370	2367	2370.5	2374	2377.4	2380.9	2384.4	2387.9	2391.3	2394.8	2398.3
380	2401.8	2405.2	2408.7	2412.2	2415.6	2419.1	2422.6	2426	2429.5	2432.9
390	2436.4	2439.9	2443.3	2446.8	2450.2	2453.7	2457.1	2460.6	2464	2467.5
400	2470.9	2474.4	2477.8	2481.3	2484.7	2488.1	2491.6	2495	2498.5	2501.9
410	2505.3	2508.8	2512.2	2515.6	2519.1	2522.5	2525.9	2529.3	2532.8	2536.2
420	2539.6	2543	2546.5	2549.9	2553.3	2556.7	2560.1	2563.5	2567	2570.4
430	2573.8	2577.2	2580.6	2584	2587.4	2590.8	2594.2	2597.6	2601	2604.4
440	2607.8	2611.2	2614.6	2618	2621.4	2624.8	2628.2	2631.6	2635	2638.4
450	2641.8	2645.2	2648.6	2652	2655.3	2658.7	2662.1	2665.5	2668.9	2672.2
460	2675.6	2679	2682.4	2685.7	2689.1	2692.5	2695.9	2699.2	2702.6	2706
470	2709.3	2712.7	2716.1	2719.4	2722.8	2726.1	2729.5	2732.9	2736.2	2739.6
480	2742.9	2746.3	2749.6	2753	2756.3	2759.7	2763	2766.4	2769.7	2773.1
490	2776.4	2779.8	2783.1	2786.4	2789.8	2793.1	2796.4	2799.8	2803.1	2806.4
500	2809.8	2813.1	2816.4	2819.8	2823.1	2826.4	2829.7	2833.1	2836.4	2839.7

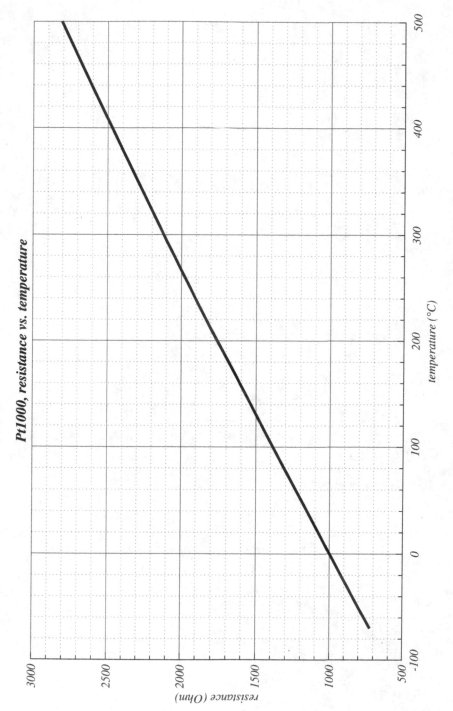

FIGURE D.1: Pt1000, resistance vs. temperature, based on data from [78]

E

Ni120

TABLE E.1: Ni120, resistance in ohm vs. temperature in °C, data [80]

°C	0	1	2	3	4	5	6	7	8	9
-80	66.60									
-70	73.10	72.45	71.80	71.15	70.50	69.85	69.20	68.55	67.90	67.25
-60	79.62	78.97	78.31	77.66	77.01	76.36	75.71	75.06	74.41	73.75
-50	86.16	85.51	84.85	84.20	83.54	82.89	82.23	81.58	80.93	80.27
-40	92.76	92.09	91.43	90.77	90.11	89.45	88.79	88.14	87.48	86.82
-30	99.41	98.74	98.07	97.41	96.74	96.07	95.41	94.74	94.08	93.42
-20	106.15	105.47	104.79	104.12	103.44	102.77	102.09	101.42	100.75	100.08
-10	113.00	112.31	111.62	110.93	110.25	109.56	108.88	108.19	107.51	106.83
0	120.00	119.29	118.59	117.88	117.18	116.48	115.78	115.09	114.39	113.70
10	127.17	127.89	128.62	129.35	130.09	130.82	131.56	132.29	133.03	133.77
20	134.52	135.26	136.01	136.76	137.51	138.26	139.02	139.78	140.54	141.30
30	142.06	142.82	143.59	144.36	145.13	145.90	146.68	147.46	148.24	149.02
40	149.80	150.59	151.37	152.16	152.95	153.75	154.54	155.34	156.14	156.94
50	157.75	158.55	159.36	160.17	160.98	161.80	162.61	163.43	164.25	165.07
60	165.90	166.73	167.56	168.39	169.22	170.06	170.90	171.74	172.58	173.42
70	174.27	175.12	175.97	176.82	177.68	178.53	179.39	180.25	181.12	181.98
80	182.85	183.72	184.59	185.46	186.34	187.22	188.10	188.98	189.87	190.75
90	191.64	192.53	193.42	194.32	195.21	196.11	197.01	197.92	198.82	199.73
100	200.64	201.55	202.47	203.38	204.30	205.22	206.14	207.07	207.99	208.92
110	209.85	210.79	211.72	212.66	213.60	214.54	215.49	216.43	217.38	218.34
120	219.29	220.25	221.20	222.16	223.13	224.09	225.06	226.03	227.00	227.97
130	228.95	229.93	230.91	231.89	232.88	233.86	234.85	235.85	236.84	237.84
140	238.84	239.84	240.84	241.85	242.85	243.86	244.88	245.89	246.91	247.93
150	248.95	249.97	251.00	252.03	253.06	254.09	255.13	256.17	257.21	258.25
160	259.30	260.34	261.39	262.45	263.50	264.56	265.62	266.69	267.75	268.82
170	269.89	270.97	272.05	273.13	274.21	275.30	276.38	277.48	278.57	279.67
180	280.77	281.87	282.98	284.09	285.20	286.32	287.44	288.56	289.69	290.82
190	291.95	293.08	294.22	295.37	296.51	297.66	298.81	299.97	301.13	302.29
200	303.45	304.62	305.80	306.97	308.15	309.34	310.52	311.72	312.91	314.11
210	315.31	316.52	317.73	318.94	320.16	321.38	322.60	323.83	325.06	326.30
220	327.54	328.78	330.03	331.28	332.53	333.79	335.05	336.32	337.59	338.87
230	340.14	341.43	342.71	344.00	345.29	346.59	347.89	349.20	350.51	351.82
240	353.14	354.46	355.79	357.12	358.45	359.79	361.13	362.47	363.82	365.17
250	366.53	367.89	369.26	370.62	372.00	373.37	374.75	376.14	377.52	378.91
260	380.31									

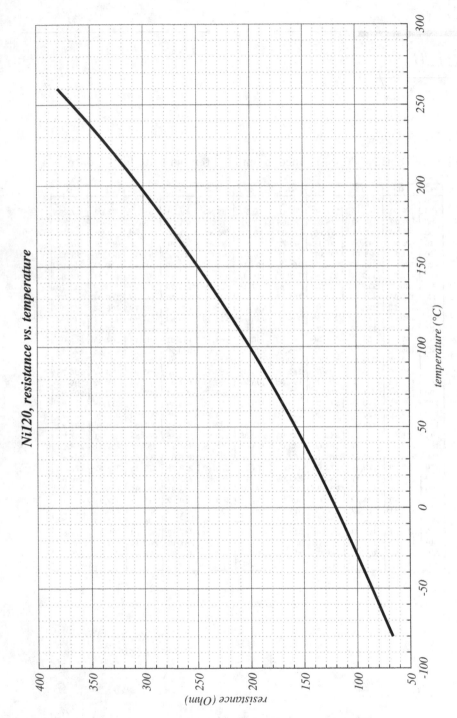

FIGURE E.1: Ni120, resistance vs. temperature, based on data from [79]

F

Cu10

TABLE F.1: Cu10, resistance in ohm vs. temperature in °C, data [82]

°C	0	1	2	3	4	5	6	7	8	9
-200	1.058									
-190	1.472	1.430	1.389	1.348	1.306	1.265	1.223	1.182	1.141	1.099
-180	1.884	1.843	1.802	1.761	1.719	1.678	1.637	1.596	1.554	1.513
-170	2.295	2.254	2.213	2.172	2.131	2.090	2.049	2.008	1.967	1.925
-160	2.705	2.664	2.623	2.582	2.541	2.500	2.459	2.418	2.377	2.336
-150	3.113	3.072	3.031	2.990	2.950	2.909	2.868	2.827	2.786	2.746
-140	3.519	3.478	3.438	3.397	3.356	3.316	3.275	3.235	3.194	3.153
-130	3.923	3.883	3.843	3.802	3.762	3.721	3.681	3.640	3.600	3.559
-120	4.327	4.286	4.246	4.206	4.165	4.125	4.085	4.045	4.004	3.964
-110	4.728	4.688	4.648	4.608	4.568	4.527	4.487	4.447	4.407	4.367
-100	5.128	5.088	5.048	5.008	4.968	4.928	4.888	4.848	4.808	4.768
-90	5.526	5.487	5.447	5.407	5.367	5.327	5.288	5.248	5.208	5.168
-80	5.923	5.884	5.844	5.804	5.765	5.725	5.685	5.646	5.606	5.566
-70	6.318	6.279	6.239	6.200	6.160	6.121	6.081	6.042	6.002	5.963
-60	6.712	6.673	6.633	6.594	6.555	6.515	6.476	6.437	6.397	6.358
-50	7.104	7.065	7.026	6.987	6.947	6.908	6.869	6.830	6.791	6.751
-40	7.490	7.452	7.413	7.374	7.336	7.297	7.259	7.220	7.181	7.143
-30	7.876	7.838	7.799	7.761	7.722	7.683	7.645	7.606	7.568	7.529
-20	8.263	8.224	8.185	8.147	8.108	8.070	8.031	7.992	7.954	7.915
-10	8.649	8.610	8.572	8.533	8.494	8.456	8.417	8.378	8.340	8.301
0	9.035	8.996	8.958	8.919	8.881	8.842	8.803	8.765	8.726	8.687
10	9.421	9.460	9.498	9.537	9.576	9.614	9.653	9.692	9.730	9.769
20	9.807	9.846	9.885	9.923	9.962	10.000	10.039	10.078	10.116	10.155
30	10.194	10.232	10.271	10.309	10.348	10.387	10.425	10.464	10.502	10.541
40	10.580	10.618	10.657	10.696	10.734	10.773	10.811	10.850	10.889	10.927
50	10.966	11.005	11.043	11.082	11.120	11.159	11.198	11.236	11.275	11.313
60	11.352	11.391	11.429	11.468	11.507	11.545	11.584	11.622	11.661	11.700
70	11.738	11.777	11.816	11.854	11.893	11.931	11.970	12.009	12.047	12.086
80	12.124	12.163	12.202	12.240	12.279	12.318	12.356	12.395	12.433	12.472
90	12.511	12.549	12.588	12.627	12.665	12.704	12.742	12.781	12.820	12.858
100	12.897	12.935	12.974	13.013	13.051	13.090	13.129	13.167	13.206	13.244
110	13.283	13.322	13.360	13.399	13.437	13.476	13.515	13.553	13.592	13.631
120	13.669	13.708	13.746	13.785	13.824	13.862	13.901	13.940	13.978	14.017
130	14.055	14.094	14.133	14.171	14.210	14.248	14.287	14.326	14.364	14.403
140	14.442	14.480	14.519	14.557	14.596	14.635	14.673	14.712	14.751	14.789
150	14.828	14.867	14.906	14.945	14.984	15.022	15.061	15.100	15.139	15.178
160	15.217	15.256	15.295	15.334	15.373	15.412	15.451	15.490	15.529	15.568
170	15.607	15.646	15.685	15.724	15.763	15.802	15.840	15.879	15.918	15.957
180	15.996	16.035	16.074	16.113	16.152	16.191	16.230	16.269	16.308	16.347
190	16.386	16.425	16.464	16.503	16.542	16.581	16.620	16.659	16.698	16.737
200	16.776	16.815	16.854	16.893	16.932	16.971	17.010	17.049	17.088	17.127
210	17.166	17.205	17.244	17.283	17.322	17.360	17.399	17.438	17.477	17.516
220	17.555	17.594	17.633	17.672	17.711	17.750	17.789	17.828	17.867	17.906
230	17.945	17.984	18.023	18.062	18.101	18.140	18.179	18.218	18.257	18.296
240	18.335	18.374	18.413	18.452	18.491	18.530	18.569	18.609	18.648	18.687
250	18.726	18.765	18.804	18.843	18.882	18.921	18.960	18.999	19.038	19.077
260	19.116									

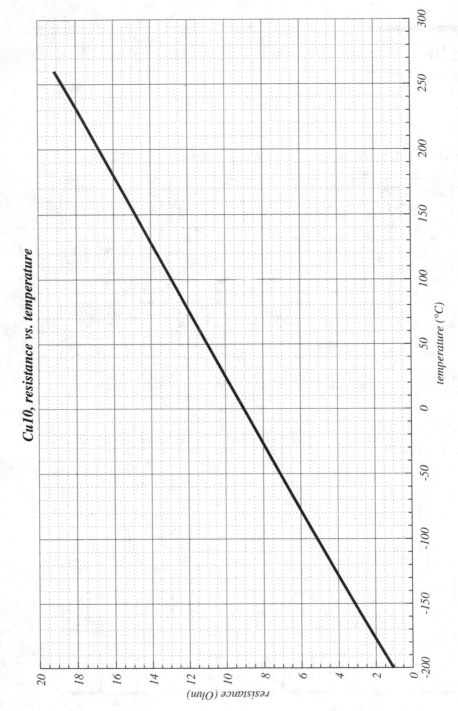

Cu10, resistance vs. temperature

FIGURE F.1: Cu10, resistance vs. temperature, based on data from [81], note that R = 10 ohm is specified for 25°C, not 0°C as other RTDs

G

Thermocouple type J

TABLE G.1: Thermocouple type J, negative temperatures, cold junction 0°C, data [83]

°C	0	-1	-2	-3	-4	-5	-6	-7	-8	-9	-10
-200	-7.89	-7.912	-7.934	-7.955	-7.976	-7.996	-8.017	-8.037	-8.057	-8.076	-8.095
-190	-7.659	-7.683	-7.707	-7.731	-7.755	-7.778	-7.801	-7.824	-7.846	-7.868	-7.89
-180	-7.403	-7.429	-7.456	-7.482	-7.508	-7.534	-7.559	-7.585	-7.61	-7.634	-7.659
-170	-7.123	-7.152	-7.181	-7.209	-7.237	-7.265	-7.293	-7.321	-7.348	-7.376	-7.403
-160	-6.821	-6.853	-6.883	-6.914	-6.944	-6.975	-7.005	-7.035	-7.064	-7.094	-7.123
-150	-6.5	-6.533	-6.566	-6.598	-6.631	-6.663	-6.695	-6.727	-6.759	-6.79	-6.821
-140	-6.159	-6.194	-6.229	-6.263	-6.298	-6.332	-6.366	-6.4	-6.433	-6.467	-6.5
-130	-5.801	-5.838	-5.874	-5.91	-5.946	-5.982	-6.018	-6.054	-6.089	-6.124	-6.159
-120	-5.426	-5.465	-5.503	-5.541	-5.578	-5.616	-5.653	-5.69	-5.727	-5.764	-5.801
-110	-5.037	-5.076	-5.116	-5.155	-5.194	-5.233	-5.272	-5.311	-5.35	-5.388	-5.426
-100	-4.633	-4.674	-4.714	-4.755	-4.796	-4.836	-4.877	-4.917	-4.957	-4.997	-5.037
-90	-4.215	-4.257	-4.3	-4.342	-4.384	-4.425	-4.467	-4.509	-4.55	-4.591	-4.633
-80	-3.786	-3.829	-3.872	-3.916	-3.959	-4.002	-4.045	-4.088	-4.13	-4.173	-4.215
-70	-3.344	-3.389	-3.434	-3.478	-3.522	-3.566	-3.61	-3.654	-3.698	-3.742	-3.786
-60	-2.893	-2.938	-2.984	-3.029	-3.075	-3.12	-3.165	-3.21	-3.255	-3.3	-3.344
-50	-2.431	-2.478	-2.524	-2.571	-2.617	-2.663	-2.709	-2.755	-2.801	-2.847	-2.893
-40	-1.961	-2.008	-2.055	-2.103	-2.15	-2.197	-2.244	-2.291	-2.338	-2.385	-2.431
-30	-1.482	-1.53	-1.578	-1.626	-1.674	-1.722	-1.77	-1.818	-1.865	-1.913	-1.961
-20	-0.995	-1.044	-1.093	-1.142	-1.19	-1.239	-1.288	-1.336	-1.385	-1.433	-1.482
-10	-0.501	-0.55	-0.6	-0.65	-0.699	-0.749	-0.798	-0.847	-0.896	-0.946	-0.995
0	0	-0.05	-0.101	-0.151	-0.201	-0.251	-0.301	-0.351	-0.401	-0.451	-0.501

TABLE G.2: Thermocouple type J, positive temperatures, cold junction 0°C, data [83]

°C	0	1	2	3	4	5	6	7	8	9	10
0	0	0.05	0.101	0.151	0.202	0.253	0.303	0.354	0.405	0.456	0.507
10	0.507	0.558	0.609	0.66	0.711	0.762	0.814	0.865	0.916	0.968	1.019
20	1.019	1.071	1.122	1.174	1.226	1.277	1.329	1.381	1.433	1.485	1.537
30	1.537	1.589	1.641	1.693	1.745	1.797	1.849	1.902	1.954	2.006	2.059
40	2.059	2.111	2.164	2.216	2.269	2.322	2.374	2.427	2.48	2.532	2.585
50	2.585	2.638	2.691	2.744	2.797	2.85	2.903	2.956	3.009	3.062	3.116
60	3.116	3.169	3.222	3.275	3.329	3.382	3.436	3.489	3.543	3.596	3.65
70	3.65	3.703	3.757	3.81	3.864	3.918	3.971	4.025	4.079	4.133	4.187
80	4.187	4.24	4.294	4.348	4.402	4.456	4.51	4.564	4.618	4.672	4.726
90	4.726	4.781	4.835	4.889	4.943	4.997	5.052	5.106	5.16	5.215	5.269
100	5.269	5.323	5.378	5.432	5.487	5.541	5.595	5.65	5.705	5.759	5.814
110	5.814	5.868	5.923	5.977	6.032	6.087	6.141	6.196	6.251	6.306	6.36
120	6.36	6.415	6.47	6.525	6.579	6.634	6.689	6.744	6.799	6.854	6.909
130	6.909	6.964	7.019	7.074	7.129	7.184	7.239	7.294	7.349	7.404	7.459
140	7.459	7.514	7.569	7.624	7.679	7.734	7.789	7.844	7.9	7.955	8.01
150	8.01	8.065	8.12	8.175	8.231	8.286	8.341	8.396	8.452	8.507	8.562

Continues on next page

TABLE G.2 – cont.

°C	0	1	2	3	4	5	6	7	8	9	10
160	8.562	8.618	8.673	8.728	8.783	8.839	8.894	8.949	9.005	9.06	9.115
170	9.115	9.171	9.226	9.282	9.337	9.392	9.448	9.503	9.559	9.614	9.669
180	9.669	9.725	9.78	9.836	9.891	9.947	10.002	10.057	10.113	10.168	10.224
190	10.224	10.279	10.335	10.39	10.446	10.501	10.557	10.612	10.668	10.723	10.779
200	10.779	10.834	10.89	10.945	11.001	11.056	11.112	11.167	11.223	11.278	11.334
210	11.334	11.389	11.445	11.501	11.556	11.612	11.667	11.723	11.778	11.834	11.889
220	11.889	11.945	12	12.056	12.111	12.167	12.222	12.278	12.334	12.389	12.445
230	12.445	12.5	12.556	12.611	12.667	12.722	12.778	12.833	12.889	12.944	13
240	13	13.056	13.111	13.167	13.222	13.278	13.333	13.389	13.444	13.5	13.555
250	13.555	13.611	13.666	13.722	13.777	13.833	13.888	13.944	13.999	14.055	14.11
260	14.11	14.166	14.221	14.277	14.332	14.388	14.443	14.499	14.554	14.609	14.665
270	14.665	14.72	14.776	14.831	14.887	14.942	14.998	15.053	15.109	15.164	15.219
280	15.219	15.275	15.33	15.386	15.441	15.496	15.552	15.607	15.663	15.718	15.773
290	15.773	15.829	15.884	15.94	15.995	16.05	16.106	16.161	16.216	16.272	16.327
300	16.327	16.383	16.438	16.493	16.549	16.604	16.659	16.715	16.77	16.825	16.881
310	16.881	16.936	16.991	17.046	17.102	17.157	17.212	17.268	17.323	17.378	17.434
320	17.434	17.489	17.544	17.599	17.655	17.71	17.765	17.82	17.876	17.931	17.986
330	17.986	18.041	18.097	18.152	18.207	18.262	18.318	18.373	18.428	18.483	18.538
340	18.538	18.594	18.649	18.704	18.759	18.814	18.87	18.925	18.98	19.035	19.09
350	19.09	19.146	19.201	19.256	19.311	19.366	19.422	19.477	19.532	19.587	19.642
360	19.642	19.697	19.753	19.808	19.863	19.918	19.973	20.028	20.083	20.139	20.194
370	20.194	20.249	20.304	20.359	20.414	20.469	20.525	20.58	20.635	20.69	20.745
380	20.745	20.8	20.855	20.911	20.966	21.021	21.076	21.131	21.186	21.241	21.297
390	21.297	21.352	21.407	21.462	21.517	21.572	21.627	21.683	21.738	21.793	21.848
400	21.848	21.903	21.958	22.014	22.069	22.124	22.179	22.234	22.289	22.345	22.4
410	22.4	22.455	22.51	22.565	22.62	22.676	22.731	22.786	22.841	22.896	22.952
420	22.952	23.007	23.062	23.117	23.172	23.228	23.283	23.338	23.393	23.449	23.504
430	23.504	23.559	23.614	23.67	23.725	23.78	23.835	23.891	23.946	24.001	24.057
440	24.057	24.112	24.167	24.223	24.278	24.333	24.389	24.444	24.499	24.555	24.61
450	24.61	24.665	24.721	24.776	24.832	24.887	24.943	24.998	25.053	25.109	25.164
460	25.164	25.22	25.275	25.331	25.386	25.442	25.497	25.553	25.608	25.664	25.72
470	25.72	25.775	25.831	25.886	25.942	25.998	26.053	26.109	26.165	26.22	26.276
480	26.276	26.332	26.387	26.443	26.499	26.555	26.61	26.666	26.722	26.778	26.834
490	26.834	26.889	26.945	27.001	27.057	27.113	27.169	27.225	27.281	27.337	27.393
500	27.393	27.449	27.505	27.561	27.617	27.673	27.729	27.785	27.841	27.897	27.953
510	27.953	28.01	28.066	28.122	28.178	28.234	28.291	28.347	28.403	28.46	28.516
520	28.516	28.572	28.629	28.685	28.741	28.798	28.854	28.911	28.967	29.024	29.08
530	29.08	29.137	29.194	29.25	29.307	29.363	29.42	29.477	29.534	29.59	29.647
540	29.647	29.704	29.761	29.818	29.874	29.931	29.988	30.045	30.102	30.159	30.216
550	30.216	30.273	30.33	30.387	30.444	30.502	30.559	30.616	30.673	30.73	30.788
560	30.788	30.845	30.902	30.96	31.017	31.074	31.132	31.189	31.247	31.304	31.362
570	31.362	31.419	31.477	31.535	31.592	31.65	31.708	31.766	31.823	31.881	31.939
580	31.939	31.997	32.055	32.113	32.171	32.229	32.287	32.345	32.403	32.461	32.519
590	32.519	32.577	32.636	32.694	32.752	32.81	32.869	32.927	32.985	33.044	33.102
600	33.102	33.161	33.219	33.278	33.337	33.395	33.454	33.513	33.571	33.63	33.689
610	33.689	33.748	33.807	33.866	33.925	33.984	34.043	34.102	34.161	34.22	34.279
620	34.279	34.338	34.397	34.457	34.516	34.575	34.635	34.694	34.754	34.813	34.873
630	34.873	34.932	34.992	35.051	35.111	35.171	35.23	35.29	35.35	35.41	35.47
640	35.47	35.53	35.59	35.65	35.71	35.77	35.83	35.89	35.95	36.01	36.071
650	36.071	36.131	36.191	36.252	36.312	36.373	36.433	36.494	36.554	36.615	36.675
660	36.675	36.736	36.797	36.858	36.918	36.979	37.04	37.101	37.162	37.223	37.284
670	37.284	37.345	37.406	37.467	37.528	37.59	37.651	37.712	37.773	37.835	37.896
680	37.896	37.958	38.019	38.081	38.142	38.204	38.265	38.327	38.389	38.45	38.512
690	38.512	38.574	38.636	38.698	38.76	38.822	38.884	38.946	39.008	39.07	39.132
700	39.132	39.194	39.256	39.318	39.381	39.443	39.505	39.568	39.63	39.693	39.755
710	39.755	39.818	39.88	39.943	40.005	40.068	40.131	40.193	40.256	40.319	40.382
720	40.382	40.445	40.508	40.57	40.633	40.696	40.759	40.822	40.886	40.949	41.012
730	41.012	41.075	41.138	41.201	41.265	41.328	41.391	41.455	41.518	41.581	41.645
740	41.645	41.708	41.772	41.835	41.899	41.962	42.026	42.09	42.153	42.217	42.281
750	42.281	42.344	42.408	42.472	42.536	42.599	42.663	42.727	42.791	42.855	42.919
760	42.919	42.983	43.047	43.111	43.175	43.239	43.303	43.367	43.431	43.495	43.559

Continues on next page

TABLE G.2 – cont.

°C	0	1	2	3	4	5	6	7	8	9	10
770	43.559	43.624	43.688	43.752	43.817	43.881	43.945	44.01	44.074	44.139	44.203
780	44.203	44.267	44.332	44.396	44.461	44.525	44.59	44.655	44.719	44.784	44.848
790	44.848	44.913	44.977	45.042	45.107	45.171	45.236	45.301	45.365	45.43	45.494
800	45.494	45.559	45.624	45.688	45.753	45.818	45.882	45.947	46.011	46.076	46.141
810	46.141	46.205	46.27	46.334	46.399	46.464	46.528	46.593	46.657	46.722	46.786
820	46.786	46.851	46.915	46.98	47.044	47.109	47.173	47.238	47.302	47.367	47.431
830	47.431	47.495	47.56	47.624	47.688	47.753	47.817	47.881	47.946	48.01	48.074
840	48.074	48.138	48.202	48.267	48.331	48.395	48.459	48.523	48.587	48.651	48.715
850	48.715	48.779	48.843	48.907	48.971	49.034	49.098	49.162	49.226	49.29	49.353
860	49.353	49.417	49.481	49.544	49.608	49.672	49.735	49.799	49.862	49.926	49.989
870	49.989	50.052	50.116	50.179	50.243	50.306	50.369	50.432	50.495	50.559	50.622
880	50.622	50.685	50.748	50.811	50.874	50.937	51	51.063	51.126	51.188	51.251
890	51.251	51.314	51.377	51.439	51.502	51.565	51.627	51.69	51.752	51.815	51.877
900	51.877	51.94	52.002	52.064	52.127	52.189	52.251	52.314	52.376	52.438	52.5
910	52.5	52.562	52.624	52.686	52.748	52.81	52.872	52.934	52.996	53.057	53.119
920	53.119	53.181	53.243	53.304	53.366	53.427	53.489	53.55	53.612	53.673	53.735
930	53.735	53.796	53.857	53.919	53.98	54.041	54.102	54.164	54.225	54.286	54.347
940	54.347	54.408	54.469	54.53	54.591	54.652	54.713	54.773	54.834	54.895	54.956
950	54.956	55.016	55.077	55.138	55.198	55.259	55.319	55.38	55.44	55.501	55.561
960	55.561	55.622	55.682	55.742	55.803	55.863	55.923	55.983	56.043	56.104	56.164
970	56.164	56.224	56.284	56.344	56.404	56.464	56.524	56.584	56.643	56.703	56.763
980	56.763	56.823	56.883	56.942	57.002	57.062	57.121	57.181	57.24	57.3	57.36
990	57.36	57.419	57.479	57.538	57.597	57.657	57.716	57.776	57.835	57.894	57.953
1000	57.953	58.013	58.072	58.131	58.19	58.249	58.309	58.368	58.427	58.486	58.545
1010	58.545	58.604	58.663	58.722	58.781	58.84	58.899	58.957	59.016	59.075	59.134
1020	59.134	59.193	59.252	59.31	59.369	59.428	59.487	59.545	59.604	59.663	59.721
1030	59.721	59.78	59.838	59.897	59.956	60.014	60.073	60.131	60.19	60.248	60.307
1040	60.307	60.365	60.423	60.482	60.54	60.599	60.657	60.715	60.774	60.832	60.89
1050	60.89	60.949	61.007	61.065	61.123	61.182	61.24	61.298	61.356	61.415	61.473
1060	61.473	61.531	61.589	61.647	61.705	61.763	61.822	61.88	61.938	61.996	62.054
1070	62.054	62.112	62.17	62.228	62.286	62.344	62.402	62.46	62.518	62.576	62.634
1080	62.634	62.692	62.75	62.808	62.866	62.924	62.982	63.04	63.098	63.156	63.214
1090	63.214	63.271	63.329	63.387	63.445	63.503	63.561	63.619	63.677	63.734	63.792
1100	63.792	63.85	63.908	63.966	64.024	64.081	64.139	64.197	64.255	64.313	64.37
1110	64.37	64.428	64.486	64.544	64.602	64.659	64.717	64.775	64.833	64.89	64.948
1120	64.948	65.006	65.064	65.121	65.179	65.237	65.295	65.352	65.41	65.468	65.525
1130	65.525	65.583	65.641	65.699	65.756	65.814	65.872	65.929	65.987	66.045	66.102
1140	66.102	66.16	66.218	66.275	66.333	66.391	66.448	66.506	66.564	66.621	66.679
1150	66.679	66.737	66.794	66.852	66.91	66.967	67.025	67.082	67.14	67.198	67.255
1160	67.255	67.313	67.37	67.428	67.486	67.543	67.601	67.658	67.716	67.773	67.831
1170	67.831	67.888	67.946	68.003	68.061	68.119	68.176	68.234	68.291	68.348	68.406
1180	68.406	68.463	68.521	68.578	68.636	68.693	68.751	68.808	68.865	68.923	68.98
1190	68.98	69.037	69.095	69.152	69.209	69.267	69.324	69.381	69.439	69.496	69.553

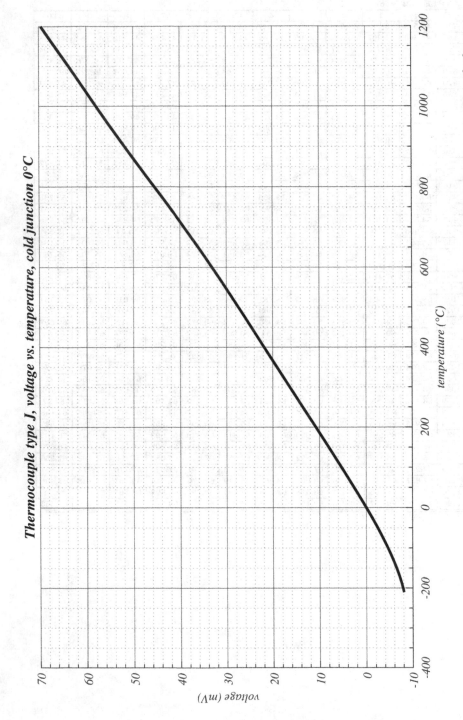

FIGURE G.1: Thermocouple type J, voltage vs. temperature, cold junction 0°C, based on data from [83]

H

Thermocouple type K

TABLE H.1: Thermocouple type K, negative temperatures, cold junction 0°C, data [83]

°C	0	-1	-2	-3	-4	-5	-6	-7	-8	-9	-10
-260	-6.441	-6.444	-6.446	-6.448	-6.45	-6.452	-6.453	-6.455	-6.456	-6.457	-6.458
-250	-6.404	-6.408	-6.413	-6.417	-6.421	-6.425	-6.429	-6.432	-6.435	-6.438	-6.441
-240	-6.344	-6.351	-6.358	-6.364	-6.37	-6.377	-6.382	-6.388	-6.393	-6.399	-6.404
-230	-6.262	-6.271	-6.28	-6.289	-6.297	-6.306	-6.314	-6.322	-6.329	-6.337	-6.344
-220	-6.158	-6.17	-6.181	-6.192	-6.202	-6.213	-6.223	-6.233	-6.243	-6.252	-6.262
-210	-6.035	-6.048	-6.061	-6.074	-6.087	-6.099	-6.111	-6.123	-6.135	-6.147	-6.158
-200	-5.891	-5.907	-5.922	-5.936	-5.951	-5.965	-5.98	-5.994	-6.007	-6.021	-6.035
-190	-5.73	-5.747	-5.763	-5.78	-5.797	-5.813	-5.829	-5.845	-5.861	-5.876	-5.891
-180	-5.55	-5.569	-5.588	-5.606	-5.624	-5.642	-5.66	-5.678	-5.695	-5.713	-5.73
-170	-5.354	-5.374	-5.395	-5.415	-5.435	-5.454	-5.474	-5.493	-5.512	-5.531	-5.55
-160	-5.141	-5.163	-5.185	-5.207	-5.228	-5.25	-5.271	-5.292	-5.313	-5.333	-5.354
-150	-4.913	-4.936	-4.96	-4.983	-5.006	-5.029	-5.052	-5.074	-5.097	-5.119	-5.141
-140	-4.669	-4.694	-4.719	-4.744	-4.768	-4.793	-4.817	-4.841	-4.865	-4.889	-4.913
-130	-4.411	-4.437	-4.463	-4.49	-4.516	-4.542	-4.567	-4.593	-4.618	-4.644	-4.669
-120	-4.138	-4.166	-4.194	-4.221	-4.249	-4.276	-4.303	-4.33	-4.357	-4.384	-4.411
-110	-3.852	-3.882	-3.911	-3.939	-3.968	-3.997	-4.025	-4.054	-4.082	-4.11	-4.138
-100	-3.554	-3.584	-3.614	-3.645	-3.675	-3.705	-3.734	-3.764	-3.794	-3.823	-3.852
-90	-3.243	-3.274	-3.306	-3.337	-3.368	-3.4	-3.431	-3.462	-3.492	-3.523	-3.554
-80	-2.92	-2.953	-2.986	-3.018	-3.05	-3.083	-3.115	-3.147	-3.179	-3.211	-3.243
-70	-2.587	-2.62	-2.654	-2.688	-2.721	-2.755	-2.788	-2.821	-2.854	-2.887	-2.92
-60	-2.243	-2.278	-2.312	-2.347	-2.382	-2.416	-2.45	-2.485	-2.519	-2.553	-2.587
-50	-1.889	-1.925	-1.961	-1.996	-2.032	-2.067	-2.103	-2.138	-2.173	-2.208	-2.243
-40	-1.527	-1.564	-1.6	-1.637	-1.673	-1.709	-1.745	-1.782	-1.818	-1.854	-1.889
-30	-1.156	-1.194	-1.231	-1.268	-1.305	-1.343	-1.38	-1.417	-1.453	-1.49	-1.527
-20	-0.778	-0.816	-0.854	-0.892	-0.93	-0.968	-1.006	-1.043	-1.081	-1.119	-1.156
-10	-0.392	-0.431	-0.47	-0.508	-0.547	-0.586	-0.624	-0.663	-0.701	-0.739	-0.778
0	0	-0.039	-0.079	-0.118	-0.157	-0.197	-0.236	-0.275	-0.314	-0.353	-0.392

TABLE H.2: Thermocouple type K, positive temperatures, cold junction 0°C, data [83]

°C	0	1	2	3	4	5	6	7	8	9	10
0	0	0.039	0.079	0.119	0.158	0.198	0.238	0.277	0.317	0.357	0.397
10	0.397	0.437	0.477	0.517	0.557	0.597	0.637	0.677	0.718	0.758	0.798
20	0.798	0.838	0.879	0.919	0.96	1	1.041	1.081	1.122	1.163	1.203
30	1.203	1.244	1.285	1.326	1.366	1.407	1.448	1.489	1.53	1.571	1.612
40	1.612	1.653	1.694	1.735	1.776	1.817	1.858	1.899	1.941	1.982	2.023
50	2.023	2.064	2.106	2.147	2.188	2.23	2.271	2.312	2.354	2.395	2.436
60	2.436	2.478	2.519	2.561	2.602	2.644	2.685	2.727	2.768	2.81	2.851
70	2.851	2.893	2.934	2.976	3.017	3.059	3.1	3.142	3.184	3.225	3.267
80	3.267	3.308	3.35	3.391	3.433	3.474	3.516	3.557	3.599	3.64	3.682
90	3.682	3.723	3.765	3.806	3.848	3.889	3.931	3.972	4.013	4.055	4.096

Continues on next page

TABLE H.2 – cont.

°C	0	1	2	3	4	5	6	7	8	9	10
100	4.096	4.138	4.179	4.22	4.262	4.303	4.344	4.385	4.427	4.468	4.509
110	4.509	4.55	4.591	4.633	4.674	4.715	4.756	4.797	4.838	4.879	4.92
120	4.92	4.961	5.002	5.043	5.084	5.124	5.165	5.206	5.247	5.288	5.328
130	5.328	5.369	5.41	5.45	5.491	5.532	5.572	5.613	5.653	5.694	5.735
140	5.735	5.775	5.815	5.856	5.896	5.937	5.977	6.017	6.058	6.098	6.138
150	6.138	6.179	6.219	6.259	6.299	6.339	6.38	6.42	6.46	6.5	6.54
160	6.54	6.58	6.62	6.66	6.701	6.741	6.781	6.821	6.861	6.901	6.941
170	6.941	6.981	7.021	7.06	7.1	7.14	7.18	7.22	7.26	7.3	7.34
180	7.34	7.38	7.42	7.46	7.5	7.54	7.579	7.619	7.659	7.699	7.739
190	7.739	7.779	7.819	7.859	7.899	7.939	7.979	8.019	8.059	8.099	8.138
200	8.138	8.178	8.218	8.258	8.298	8.338	8.378	8.418	8.458	8.499	8.539
210	8.539	8.579	8.619	8.659	8.699	8.739	8.779	8.819	8.86	8.9	8.94
220	8.94	8.98	9.02	9.061	9.101	9.141	9.181	9.222	9.262	9.302	9.343
230	9.343	9.383	9.423	9.464	9.504	9.545	9.585	9.626	9.666	9.707	9.747
240	9.747	9.788	9.828	9.869	9.909	9.95	9.991	10.031	10.072	10.113	10.153
250	10.153	10.194	10.235	10.276	10.316	10.357	10.398	10.439	10.48	10.52	10.561
260	10.561	10.602	10.643	10.684	10.725	10.766	10.807	10.848	10.889	10.93	10.971
270	10.971	11.012	11.053	11.094	11.135	11.176	11.217	11.259	11.3	11.341	11.382
280	11.382	11.423	11.465	11.506	11.547	11.588	11.63	11.671	11.712	11.753	11.795
290	11.795	11.836	11.877	11.919	11.96	12.001	12.043	12.084	12.126	12.167	12.209
300	12.209	12.25	12.291	12.333	12.374	12.416	12.457	12.499	12.54	12.582	12.624
310	12.624	12.665	12.707	12.748	12.79	12.831	12.873	12.915	12.956	12.998	13.04
320	13.04	13.081	13.123	13.165	13.206	13.248	13.29	13.331	13.373	13.415	13.457
330	13.457	13.498	13.54	13.582	13.624	13.665	13.707	13.749	13.791	13.833	13.874
340	13.874	13.916	13.958	14	14.042	14.084	14.126	14.167	14.209	14.251	14.293
350	14.293	14.335	14.377	14.419	14.461	14.503	14.545	14.587	14.629	14.671	14.713
360	14.713	14.755	14.797	14.839	14.881	14.923	14.965	15.007	15.049	15.091	15.133
370	15.133	15.175	15.217	15.259	15.301	15.343	15.385	15.427	15.469	15.511	15.554
380	15.554	15.596	15.638	15.68	15.722	15.764	15.806	15.849	15.891	15.933	15.975
390	15.975	16.017	16.059	16.102	16.144	16.186	16.228	16.27	16.313	16.355	16.397
400	16.397	16.439	16.482	16.524	16.566	16.608	16.651	16.693	16.735	16.778	16.82
410	16.82	16.862	16.904	16.947	16.989	17.031	17.074	17.116	17.158	17.201	17.243
420	17.243	17.285	17.328	17.37	17.413	17.455	17.497	17.54	17.582	17.624	17.667
430	17.667	17.709	17.752	17.794	17.837	17.879	17.921	17.964	18.006	18.049	18.091
440	18.091	18.134	18.176	18.218	18.261	18.303	18.346	18.388	18.431	18.473	18.516
450	18.516	18.558	18.601	18.643	18.686	18.728	18.771	18.813	18.856	18.898	18.941
460	18.941	18.983	19.026	19.068	19.111	19.154	19.196	19.239	19.281	19.324	19.366
470	19.366	19.409	19.451	19.494	19.537	19.579	19.622	19.664	19.707	19.75	19.792
480	19.792	19.835	19.877	19.92	19.962	20.005	20.048	20.09	20.133	20.175	20.218
490	20.218	20.261	20.303	20.346	20.389	20.431	20.474	20.516	20.559	20.602	20.644
500	20.644	20.687	20.73	20.772	20.815	20.857	20.9	20.943	20.985	21.028	21.071
510	21.071	21.113	21.156	21.199	21.241	21.284	21.326	21.369	21.412	21.454	21.497
520	21.497	21.54	21.582	21.625	21.668	21.71	21.753	21.796	21.838	21.881	21.924
530	21.924	21.966	22.009	22.052	22.094	22.137	22.179	22.222	22.265	22.307	22.35
540	22.35	22.393	22.435	22.478	22.521	22.563	22.606	22.649	22.691	22.734	22.776
550	22.776	22.819	22.862	22.904	22.947	22.99	23.032	23.075	23.117	23.16	23.203
560	23.203	23.245	23.288	23.331	23.373	23.416	23.458	23.501	23.544	23.586	23.629
570	23.629	23.671	23.714	23.757	23.799	23.842	23.884	23.927	23.97	24.012	24.055
580	24.055	24.097	24.14	24.182	24.225	24.267	24.31	24.353	24.395	24.438	24.48
590	24.48	24.523	24.565	24.608	24.65	24.693	24.735	24.778	24.82	24.863	24.905
600	24.905	24.948	24.99	25.033	25.075	25.118	25.16	25.203	25.245	25.288	25.33
610	25.33	25.373	25.415	25.458	25.5	25.543	25.585	25.627	25.67	25.712	25.755
620	25.755	25.797	25.84	25.882	25.924	25.967	26.009	26.052	26.094	26.136	26.179
630	26.179	26.221	26.263	26.306	26.348	26.39	26.433	26.475	26.517	26.56	26.602
640	26.602	26.644	26.687	26.729	26.771	26.814	26.856	26.898	26.94	26.983	27.025
650	27.025	27.067	27.109	27.152	27.194	27.236	27.278	27.32	27.363	27.405	27.447
660	27.447	27.489	27.531	27.574	27.616	27.658	27.7	27.742	27.784	27.826	27.869
670	27.869	27.911	27.953	27.995	28.037	28.079	28.121	28.163	28.205	28.247	28.289
680	28.289	28.332	28.374	28.416	28.458	28.5	28.542	28.584	28.626	28.668	28.71
690	28.71	28.752	28.794	28.835	28.877	28.919	28.961	29.003	29.045	29.087	29.129
700	29.129	29.171	29.213	29.255	29.297	29.338	29.38	29.422	29.464	29.506	29.548

Continues on next page

TABLE H.2 – cont.

°C	0	1	2	3	4	5	6	7	8	9	10
710	29.548	29.589	29.631	29.673	29.715	29.757	29.798	29.84	29.882	29.924	29.965
720	29.965	30.007	30.049	30.09	30.132	30.174	30.216	30.257	30.299	30.341	30.382
730	30.382	30.424	30.466	30.507	30.549	30.59	30.632	30.674	30.715	30.757	30.798
740	30.798	30.84	30.881	30.923	30.964	31.006	31.047	31.089	31.13	31.172	31.213
750	31.213	31.255	31.296	31.338	31.379	31.421	31.462	31.504	31.545	31.586	31.628
760	31.628	31.669	31.71	31.752	31.793	31.834	31.876	31.917	31.958	32	32.041
770	32.041	32.082	32.124	32.165	32.206	32.247	32.289	32.33	32.371	32.412	32.453
780	32.453	32.495	32.536	32.577	32.618	32.659	32.7	32.742	32.783	32.824	32.865
790	32.865	32.906	32.947	32.988	33.029	33.07	33.111	33.152	33.193	33.234	33.275
800	33.275	33.316	33.357	33.398	33.439	33.48	33.521	33.562	33.603	33.644	33.685
810	33.685	33.726	33.767	33.808	33.848	33.889	33.93	33.971	34.012	34.053	34.093
820	34.093	34.134	34.175	34.216	34.257	34.297	34.338	34.379	34.42	34.46	34.501
830	34.501	34.542	34.582	34.623	34.664	34.704	34.745	34.786	34.826	34.867	34.908
840	34.908	34.948	34.989	35.029	35.07	35.11	35.151	35.192	35.232	35.273	35.313
850	35.313	35.354	35.394	35.435	35.475	35.516	35.556	35.596	35.637	35.677	35.718
860	35.718	35.758	35.798	35.839	35.879	35.92	35.96	36	36.041	36.081	36.121
870	36.121	36.162	36.202	36.242	36.282	36.323	36.363	36.403	36.443	36.484	36.524
880	36.524	36.564	36.604	36.644	36.685	36.725	36.765	36.805	36.845	36.885	36.925
890	36.925	36.965	37.006	37.046	37.086	37.126	37.166	37.206	37.246	37.286	37.326
900	37.326	37.366	37.406	37.446	37.486	37.526	37.566	37.606	37.646	37.686	37.725
910	37.725	37.765	37.805	37.845	37.885	37.925	37.965	38.005	38.044	38.084	38.124
920	38.124	38.164	38.204	38.243	38.283	38.323	38.363	38.402	38.442	38.482	38.522
930	38.522	38.561	38.601	38.641	38.68	38.72	38.76	38.799	38.839	38.878	38.918
940	38.918	38.958	38.997	39.037	39.076	39.116	39.155	39.195	39.235	39.274	39.314
950	39.314	39.353	39.393	39.432	39.471	39.511	39.55	39.59	39.629	39.669	39.708
960	39.708	39.747	39.787	39.826	39.866	39.905	39.944	39.984	40.023	40.062	40.101
970	40.101	40.141	40.18	40.219	40.259	40.298	40.337	40.376	40.415	40.455	40.494
980	40.494	40.533	40.572	40.611	40.651	40.69	40.729	40.768	40.807	40.846	40.885
990	40.885	40.924	40.963	41.002	41.042	41.081	41.12	41.159	41.198	41.237	41.276
1000	41.276	41.315	41.354	41.393	41.431	41.47	41.509	41.548	41.587	41.626	41.665
1010	41.665	41.704	41.743	41.781	41.82	41.859	41.898	41.937	41.976	42.014	42.053
1020	42.053	42.092	42.131	42.169	42.208	42.247	42.286	42.324	42.363	42.402	42.44
1030	42.44	42.479	42.518	42.556	42.595	42.633	42.672	42.711	42.749	42.788	42.826
1040	42.826	42.865	42.903	42.942	42.98	43.019	43.057	43.096	43.134	43.173	43.211
1050	43.211	43.25	43.288	43.327	43.365	43.403	43.442	43.48	43.518	43.557	43.595
1060	43.595	43.633	43.672	43.71	43.748	43.787	43.825	43.863	43.901	43.94	43.978
1070	43.978	44.016	44.054	44.092	44.13	44.169	44.207	44.245	44.283	44.321	44.359
1080	44.359	44.397	44.435	44.473	44.512	44.55	44.588	44.626	44.664	44.702	44.74
1090	44.74	44.778	44.816	44.853	44.891	44.929	44.967	45.005	45.043	45.081	45.119
1100	45.119	45.157	45.194	45.232	45.27	45.308	45.346	45.383	45.421	45.459	45.497
1110	45.497	45.534	45.572	45.61	45.647	45.685	45.723	45.76	45.798	45.836	45.873
1120	45.873	45.911	45.948	45.986	46.024	46.061	46.099	46.136	46.174	46.211	46.249
1130	46.249	46.286	46.324	46.361	46.398	46.436	46.473	46.511	46.548	46.585	46.623
1140	46.623	46.66	46.697	46.735	46.772	46.809	46.847	46.884	46.921	46.958	46.995
1150	46.995	47.033	47.07	47.107	47.144	47.181	47.218	47.256	47.293	47.33	47.367
1160	47.367	47.404	47.441	47.478	47.515	47.552	47.589	47.626	47.663	47.7	47.737
1170	47.737	47.774	47.811	47.848	47.884	47.921	47.958	47.995	48.032	48.069	48.105
1180	48.105	48.142	48.179	48.216	48.252	48.289	48.326	48.363	48.399	48.436	48.473
1190	48.473	48.509	48.546	48.582	48.619	48.656	48.692	48.729	48.765	48.802	48.838
1200	48.838	48.875	48.911	48.948	48.984	49.021	49.057	49.093	49.13	49.166	49.202
1210	49.202	49.239	49.275	49.311	49.348	49.384	49.42	49.456	49.493	49.529	49.565
1220	49.565	49.601	49.637	49.674	49.71	49.746	49.782	49.818	49.854	49.89	49.926
1230	49.926	49.962	49.998	50.034	50.07	50.106	50.142	50.178	50.214	50.25	50.286
1240	50.286	50.322	50.358	50.393	50.429	50.465	50.501	50.537	50.572	50.608	50.644
1250	50.644	50.68	50.715	50.751	50.787	50.822	50.858	50.894	50.929	50.965	51
1260	51	51.036	51.071	51.107	51.142	51.178	51.213	51.249	51.284	51.32	51.355
1270	51.355	51.391	51.426	51.461	51.497	51.532	51.567	51.603	51.638	51.673	51.708
1280	51.708	51.744	51.779	51.814	51.849	51.885	51.92	51.955	51.99	52.025	52.06
1290	52.06	52.095	52.13	52.165	52.2	52.235	52.27	52.305	52.34	52.375	52.41
1300	52.41	52.445	52.48	52.515	52.55	52.585	52.62	52.654	52.689	52.724	52.759
1310	52.759	52.794	52.828	52.863	52.898	52.932	52.967	53.002	53.037	53.071	53.106

Continues on next page

TABLE H.2 – cont.

°C	0	1	2	3	4	5	6	7	8	9	10
1320	53.106	53.14	53.175	53.21	53.244	53.279	53.313	53.348	53.382	53.417	53.451
1330	53.451	53.486	53.52	53.555	53.589	53.623	53.658	53.692	53.727	53.761	53.795
1340	53.795	53.83	53.864	53.898	53.932	53.967	54.001	54.035	54.069	54.104	54.138
1350	54.138	54.172	54.206	54.24	54.274	54.308	54.343	54.377	54.411	54.445	54.479
1360	54.479	54.513	54.547	54.581	54.615	54.649	54.683	54.717	54.751	54.785	54.819

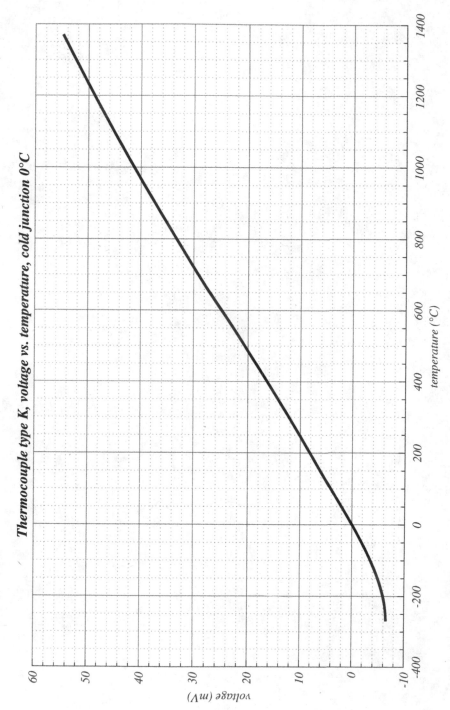

FIGURE H.1: Thermocouple type K, voltage vs. temperature, cold junction 0°C, based on data from [83]

I

Thermocouple type T

TABLE I.1: Thermocouple type T, negative temperatures, cold junction 0°C, data [83]

°C	0	-1	-2	-3	-4	-5	-6	-7	-8	-9	-10
-270	-6.258										
-260	-6.232	-6.236	-6.239	-6.242	-6.245	-6.248	-6.251	-6.253	-6.255	-6.256	-6.258
-250	-6.18	-6.187	-6.193	-6.198	-6.204	-6.209	-6.214	-6.219	-6.223	-6.228	-6.232
-240	-6.105	-6.114	-6.122	-6.13	-6.138	-6.146	-6.153	-6.16	-6.167	-6.174	-6.18
-230	-6.007	-6.017	-6.028	-6.038	-6.049	-6.059	-6.068	-6.078	-6.087	-6.096	-6.105
-220	-5.888	-5.901	-5.914	-5.926	-5.938	-5.95	-5.962	-5.973	-5.985	-5.996	-6.007
-210	-5.753	-5.767	-5.782	-5.795	-5.809	-5.823	-5.836	-5.85	-5.863	-5.876	-5.888
-200	-5.603	-5.619	-5.634	-5.65	-5.665	-5.68	-5.695	-5.71	-5.724	-5.739	-5.753
-190	-5.439	-5.456	-5.473	-5.489	-5.506	-5.523	-5.539	-5.555	-5.571	-5.587	-5.603
-180	-5.261	-5.279	-5.297	-5.316	-5.334	-5.351	-5.369	-5.387	-5.404	-5.421	-5.439
-170	-5.07	-5.089	-5.109	-5.128	-5.148	-5.167	-5.186	-5.205	-5.224	-5.242	-5.261
-160	-4.865	-4.886	-4.907	-4.928	-4.949	-4.969	-4.989	-5.01	-5.03	-5.05	-5.07
-150	-4.648	-4.671	-4.693	-4.715	-4.737	-4.759	-4.78	-4.802	-4.823	-4.844	-4.865
-140	-4.419	-4.443	-4.466	-4.489	-4.512	-4.535	-4.558	-4.581	-4.604	-4.626	-4.648
-130	-4.177	-4.202	-4.226	-4.251	-4.275	-4.3	-4.324	-4.348	-4.372	-4.395	-4.419
-120	-3.923	-3.949	-3.975	-4	-4.026	-4.052	-4.077	-4.102	-4.127	-4.152	-4.177
-110	-3.657	-3.684	-3.711	-3.738	-3.765	-3.791	-3.818	-3.844	-3.871	-3.897	-3.923
-100	-3.379	-3.407	-3.435	-3.463	-3.491	-3.519	-3.547	-3.574	-3.602	-3.629	-3.657
-90	-3.089	-3.118	-3.148	-3.177	-3.206	-3.235	-3.264	-3.293	-3.322	-3.35	-3.379
-80	-2.788	-2.818	-2.849	-2.879	-2.91	-2.94	-2.97	-3	-3.03	-3.059	-3.089
-70	-2.476	-2.507	-2.539	-2.571	-2.602	-2.633	-2.664	-2.695	-2.726	-2.757	-2.788
-60	-2.153	-2.186	-2.218	-2.251	-2.283	-2.316	-2.348	-2.38	-2.412	-2.444	-2.476
-50	-1.819	-1.853	-1.887	-1.92	-1.954	-1.987	-2.021	-2.054	-2.087	-2.12	-2.153
-40	-1.475	-1.51	-1.545	-1.579	-1.614	-1.648	-1.683	-1.717	-1.751	-1.785	-1.819
-30	-1.121	-1.157	-1.192	-1.228	-1.264	-1.299	-1.335	-1.37	-1.405	-1.44	-1.475
-20	-0.757	-0.794	-0.83	-0.867	-0.904	-0.94	-0.976	-1.013	-1.049	-1.085	-1.121
-10	-0.383	-0.421	-0.459	-0.496	-0.534	-0.571	-0.608	-0.646	-0.683	-0.72	-0.757
0	0	-0.039	-0.077	-0.116	-0.154	-0.193	-0.231	-0.269	-0.307	-0.345	-0.383

TABLE I.2: Thermocouple type T, positive temperatures, cold junction 0°C, data [83]

°C	0	1	2	3	4	5	6	7	8	9	10
0	0	0.039	0.078	0.117	0.156	0.195	0.234	0.273	0.312	0.352	0.391
10	0.391	0.431	0.47	0.51	0.549	0.589	0.629	0.669	0.709	0.749	0.79
20	0.79	0.83	0.87	0.911	0.951	0.992	1.033	1.074	1.114	1.155	1.196
30	1.196	1.238	1.279	1.32	1.362	1.403	1.445	1.486	1.528	1.57	1.612
40	1.612	1.654	1.696	1.738	1.78	1.823	1.865	1.908	1.95	1.993	2.036
50	2.036	2.079	2.122	2.165	2.208	2.251	2.294	2.338	2.381	2.425	2.468
60	2.468	2.512	2.556	2.6	2.643	2.687	2.732	2.776	2.82	2.864	2.909
70	2.909	2.953	2.998	3.043	3.087	3.132	3.177	3.222	3.267	3.312	3.358
80	3.358	3.403	3.448	3.494	3.539	3.585	3.631	3.677	3.722	3.768	3.814
90	3.814	3.86	3.907	3.953	3.999	4.046	4.092	4.138	4.185	4.232	4.279
100	4.279	4.325	4.372	4.419	4.466	4.513	4.561	4.608	4.655	4.702	4.75

Continues on next page

TABLE I.2 – cont.

°C	0	1	2	3	4	5	6	7	8	9	10
110	4.75	4.798	4.845	4.893	4.941	4.988	5.036	5.084	5.132	5.18	5.228
120	5.228	5.277	5.325	5.373	5.422	5.47	5.519	5.567	5.616	5.665	5.714
130	5.714	5.763	5.812	5.861	5.91	5.959	6.008	6.057	6.107	6.156	6.206
140	6.206	6.255	6.305	6.355	6.404	6.454	6.504	6.554	6.604	6.654	6.704
150	6.704	6.754	6.805	6.855	6.905	6.956	7.006	7.057	7.107	7.158	7.209
160	7.209	7.26	7.31	7.361	7.412	7.463	7.515	7.566	7.617	7.668	7.72
170	7.72	7.771	7.823	7.874	7.926	7.977	8.029	8.081	8.133	8.185	8.237
180	8.237	8.289	8.341	8.393	8.445	8.497	8.55	8.602	8.654	8.707	8.759
190	8.759	8.812	8.865	8.917	8.97	9.023	9.076	9.129	9.182	9.235	9.288
200	9.288	9.341	9.395	9.448	9.501	9.555	9.608	9.662	9.715	9.769	9.822
210	9.822	9.876	9.93	9.984	10.038	10.092	10.146	10.2	10.254	10.308	10.362
220	10.362	10.417	10.471	10.525	10.58	10.634	10.689	10.743	10.798	10.853	10.907
230	10.907	10.962	11.017	11.072	11.127	11.182	11.237	11.292	11.347	11.403	11.458
240	11.458	11.513	11.569	11.624	11.68	11.735	11.791	11.846	11.902	11.958	12.013
250	12.013	12.069	12.125	12.181	12.237	12.293	12.349	12.405	12.461	12.518	12.574
260	12.574	12.63	12.687	12.743	12.799	12.856	12.912	12.969	13.026	13.082	13.139
270	13.139	13.196	13.253	13.31	13.366	13.423	13.48	13.537	13.595	13.652	13.709
280	13.709	13.766	13.823	13.881	13.938	13.995	14.053	14.11	14.168	14.226	14.283
290	14.283	14.341	14.399	14.456	14.514	14.572	14.63	14.688	14.746	14.804	14.862
300	14.862	14.92	14.978	15.036	15.095	15.153	15.211	15.27	15.328	15.386	15.445
310	15.445	15.503	15.562	15.621	15.679	15.738	15.797	15.856	15.914	15.973	16.032
320	16.032	16.091	16.15	16.209	16.268	16.327	16.387	16.446	16.505	16.564	16.624
330	16.624	16.683	16.742	16.802	16.861	16.921	16.98	17.04	17.1	17.159	17.219
340	17.219	17.279	17.339	17.399	17.458	17.518	17.578	17.638	17.698	17.759	17.819
350	17.819	17.870	17.939	17.999	18.06	18.12	18.18	18.241	18.301	18.362	18.422
360	18.422	18.483	18.543	18.604	18.665	18.725	18.786	18.847	18.908	18.969	19.03
370	19.03	19.091	19.152	19.213	19.274	19.335	19.396	19.457	19.518	19.579	19.641
380	19.641	19.702	19.763	19.825	19.886	19.947	20.009	20.07	20.132	20.193	20.255
390	20.255	20.317	20.378	20.44	20.502	20.563	20.625	20.687	20.748	20.81	20.872
400	20.872										

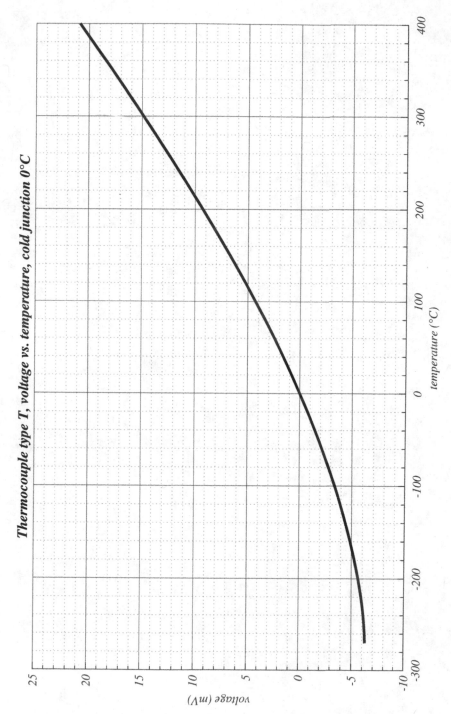

FIGURE I.1: Thermocouple type T, voltage vs. temperature, cold junction 0°C, based on data from [83]

J

Thermocouple type E

TABLE J.1: Thermocouple type E, negative temperatures, cold junction 0°C, data [83]

°C	0	-1	-2	-3	-4	-5	-6	-7	-8	-9	-10
-270	-9.835										
-260	-9.797	-9.802	-9.808	-9.813	-9.817	-9.821	-9.825	-9.828	-9.831	-9.833	-9.835
-250	-9.718	-9.728	-9.737	-9.746	-9.754	-9.762	-9.77	-9.777	-9.784	-9.79	-9.797
-240	-9.604	-9.617	-9.63	-9.642	-9.654	-9.666	-9.677	-9.688	-9.698	-9.709	-9.718
-230	-9.455	-9.471	-9.487	-9.503	-9.519	-9.534	-9.548	-9.563	-9.577	-9.591	-9.604
-220	-9.274	-9.293	-9.313	-9.331	-9.35	-9.368	-9.386	-9.404	-9.421	-9.438	-9.455
-210	-9.063	-9.085	-9.107	-9.129	-9.151	-9.172	-9.193	-9.214	-9.234	-9.254	-9.274
-200	-8.825	-8.85	-8.874	-8.899	-8.923	-8.947	-8.971	-8.994	-9.017	-9.04	-9.063
-190	-8.561	-8.588	-8.616	-8.643	-8.669	-8.696	-8.722	-8.748	-8.774	-8.799	-8.825
-180	-8.273	-8.303	-8.333	-8.362	-8.391	-8.42	-8.449	-8.477	-8.505	-8.533	-8.561
-170	-7.963	-7.995	-8.027	-8.059	-8.09	-8.121	-8.152	-8.183	-8.213	-8.243	-8.273
-160	-7.632	-7.666	-7.7	-7.733	-7.767	-7.8	-7.833	-7.866	-7.899	-7.931	-7.963
-150	-7.279	-7.315	-7.351	-7.387	-7.423	-7.458	-7.493	-7.528	-7.563	-7.597	-7.632
-140	-6.907	-6.945	-6.983	-7.021	-7.058	-7.096	-7.133	-7.17	-7.206	-7.243	-7.279
-130	-6.516	-6.556	-6.596	-6.636	-6.675	-6.714	-6.753	-6.792	-6.831	-6.869	-6.907
-120	-6.107	-6.149	-6.191	-6.232	-6.273	-6.314	-6.355	-6.396	-6.436	-6.476	-6.516
-110	-5.681	-5.724	-5.767	-5.81	-5.853	-5.896	-5.939	-5.981	-6.023	-6.065	-6.107
-100	-5.237	-5.282	-5.327	-5.372	-5.417	-5.461	-5.505	-5.549	-5.593	-5.637	-5.681
-90	-4.777	-4.824	-4.871	-4.917	-4.963	-5.009	-5.055	-5.101	-5.147	-5.192	-5.237
-80	-4.302	-4.35	-4.398	-4.446	-4.494	-4.542	-4.589	-4.636	-4.684	-4.731	-4.777
-70	-3.811	-3.861	-3.911	-3.96	-4.009	-4.058	-4.107	-4.156	-4.205	-4.254	-4.302
-60	-3.306	-3.357	-3.408	-3.459	-3.51	-3.561	-3.611	-3.661	-3.711	-3.761	-3.811
-50	-2.787	-2.84	-2.892	-2.944	-2.996	-3.048	-3.1	-3.152	-3.204	-3.255	-3.306
-40	-2.255	-2.309	-2.362	-2.416	-2.469	-2.523	-2.576	-2.629	-2.682	-2.735	-2.787
-30	-1.709	-1.765	-1.82	-1.874	-1.929	-1.984	-2.038	-2.093	-2.147	-2.201	-2.255
-20	-1.152	-1.208	-1.264	-1.32	-1.376	-1.432	-1.488	-1.543	-1.599	-1.654	-1.709
-10	-0.582	-0.639	-0.697	-0.754	-0.811	-0.868	-0.925	-0.982	-1.039	-1.095	-1.152
0	0	-0.059	-0.117	-0.176	-0.234	-0.292	-0.35	-0.408	-0.466	-0.524	-0.582

TABLE J.2: Thermocouple type E, positive temperatures, cold junction 0°C, data [83]

°C	0	1	2	3	4	5	6	7	8	9	10
0	0	0.059	0.118	0.176	0.235	0.294	0.354	0.413	0.472	0.532	0.591
10	0.591	0.651	0.711	0.77	0.83	0.89	0.95	1.01	1.071	1.131	1.192
20	1.192	1.252	1.313	1.373	1.434	1.495	1.556	1.617	1.678	1.74	1.801
30	1.801	1.862	1.924	1.986	2.047	2.109	2.171	2.233	2.295	2.357	2.42
40	2.42	2.482	2.545	2.607	2.67	2.733	2.795	2.858	2.921	2.984	3.048
50	3.048	3.111	3.174	3.238	3.301	3.365	3.429	3.492	3.556	3.62	3.685
60	3.685	3.749	3.813	3.877	3.942	4.006	4.071	4.136	4.2	4.265	4.33
70	4.33	4.395	4.46	4.526	4.591	4.656	4.722	4.788	4.853	4.919	4.985
80	4.985	5.051	5.117	5.183	5.249	5.315	5.382	5.448	5.514	5.581	5.648
90	5.648	5.714	5.781	5.848	5.915	5.982	6.049	6.117	6.184	6.251	6.319
100	6.319	6.386	6.454	6.522	6.59	6.658	6.725	6.794	6.862	6.93	6.998

Continues on next page

TABLE J.2 – cont.

°C	0	1	2	3	4	5	6	7	8	9	10
110	6.998	7.066	7.135	7.203	7.272	7.341	7.409	7.478	7.547	7.616	7.685
120	7.685	7.754	7.823	7.892	7.962	8.031	8.101	8.17	8.24	8.309	8.379
130	8.379	8.449	8.519	8.589	8.659	8.729	8.799	8.869	8.94	9.01	9.081
140	9.081	9.151	9.222	9.292	9.363	9.434	9.505	9.576	9.647	9.718	9.789
150	9.789	9.86	9.931	10.003	10.074	10.145	10.217	10.288	10.36	10.432	10.503
160	10.503	10.575	10.647	10.719	10.791	10.863	10.935	11.007	11.08	11.152	11.224
170	11.224	11.297	11.369	11.442	11.514	11.587	11.66	11.733	11.805	11.878	11.951
180	11.951	12.024	12.097	12.17	12.243	12.317	12.39	12.463	12.537	12.61	12.684
190	12.684	12.757	12.831	12.904	12.978	13.052	13.126	13.199	13.273	13.347	13.421
200	13.421	13.495	13.569	13.644	13.718	13.792	13.866	13.941	14.015	14.09	14.164
210	14.164	14.239	14.313	14.388	14.463	14.537	14.612	14.687	14.762	14.837	14.912
220	14.912	14.987	15.062	15.137	15.212	15.287	15.362	15.438	15.513	15.588	15.664
230	15.664	15.739	15.815	15.89	15.966	16.041	16.117	16.193	16.269	16.344	16.42
240	16.42	16.496	16.572	16.648	16.724	16.8	16.876	16.952	17.028	17.104	17.181
250	17.181	17.257	17.333	17.409	17.486	17.562	17.639	17.715	17.792	17.868	17.945
260	17.945	18.021	18.098	18.175	18.252	18.328	18.405	18.482	18.559	18.636	18.713
270	18.713	18.79	18.867	18.944	19.021	19.098	19.175	19.252	19.33	19.407	19.484
280	19.484	19.561	19.639	19.716	19.794	19.871	19.948	20.026	20.103	20.181	20.259
290	20.259	20.336	20.414	20.492	20.569	20.647	20.725	20.803	20.88	20.958	21.036
300	21.036	21.114	21.192	21.27	21.348	21.426	21.504	21.582	21.66	21.739	21.817
310	21.817	21.895	21.973	22.051	22.13	22.208	22.286	22.365	22.443	22.522	22.6
320	22.6	22.678	22.757	22.835	22.914	22.993	23.071	23.15	23.228	23.307	23.386
330	23.386	23.464	23.543	23.622	23.701	23.78	23.858	23.937	24.016	24.095	24.174
340	24.174	24.253	24.332	24.411	24.49	24.569	24.648	24.727	24.806	24.885	24.964
350	24.964	25.044	25.123	25.202	25.281	25.36	25.44	25.519	25.598	25.678	25.757
360	25.757	25.836	25.916	25.995	26.075	26.154	26.233	26.313	26.392	26.472	26.552
370	26.552	26.631	26.711	26.79	26.87	26.95	27.029	27.109	27.189	27.268	27.348
380	27.348	27.428	27.507	27.587	27.667	27.747	27.827	27.907	27.986	28.066	28.146
390	28.146	28.226	28.306	28.386	28.466	28.546	28.626	28.706	28.786	28.866	28.946
400	28.946	29.026	29.106	29.186	29.266	29.346	29.427	29.507	29.587	29.667	29.747
410	29.747	29.827	29.908	29.988	30.068	30.148	30.229	30.309	30.389	30.47	30.55
420	30.55	30.63	30.711	30.791	30.871	30.952	31.032	31.112	31.193	31.273	31.354
430	31.354	31.434	31.515	31.595	31.676	31.756	31.837	31.917	31.998	32.078	32.159
440	32.159	32.239	32.32	32.4	32.481	32.562	32.642	32.723	32.803	32.884	32.965
450	32.965	33.045	33.126	33.207	33.287	33.368	33.449	33.529	33.61	33.691	33.772
460	33.772	33.852	33.933	34.014	34.095	34.175	34.256	34.337	34.418	34.498	34.579
470	34.579	34.66	34.741	34.822	34.902	34.983	35.064	35.145	35.226	35.307	35.387
480	35.387	35.468	35.549	35.63	35.711	35.792	35.873	35.954	36.034	36.115	36.196
490	36.196	36.277	36.358	36.439	36.52	36.601	36.682	36.763	36.843	36.924	37.005
500	37.005	37.086	37.167	37.248	37.329	37.41	37.491	37.572	37.653	37.734	37.815
510	37.815	37.896	37.977	38.058	38.139	38.22	38.3	38.381	38.462	38.543	38.624
520	38.624	38.705	38.786	38.867	38.948	39.029	39.11	39.191	39.272	39.353	39.434
530	39.434	39.515	39.596	39.677	39.758	39.839	39.92	40.001	40.082	40.163	40.243
540	40.243	40.324	40.405	40.486	40.567	40.648	40.729	40.81	40.891	40.972	41.053
550	41.053	41.134	41.215	41.296	41.377	41.457	41.538	41.619	41.7	41.781	41.862
560	41.862	41.943	42.024	42.105	42.185	42.266	42.347	42.428	42.509	42.59	42.671
570	42.671	42.751	42.832	42.913	42.994	43.075	43.156	43.236	43.317	43.398	43.479
580	43.479	43.56	43.64	43.721	43.802	43.883	43.963	44.044	44.125	44.206	44.286
590	44.286	44.367	44.448	44.529	44.609	44.69	44.771	44.851	44.932	45.013	45.093
600	45.093	45.174	45.255	45.335	45.416	45.497	45.577	45.658	45.738	45.819	45.9
610	45.9	45.98	46.061	46.141	46.222	46.302	46.383	46.463	46.544	46.624	46.705
620	46.705	46.785	46.866	46.946	47.027	47.107	47.188	47.268	47.349	47.429	47.509
630	47.509	47.59	47.67	47.751	47.831	47.911	47.992	48.072	48.152	48.233	48.313
640	48.313	48.393	48.474	48.554	48.634	48.715	48.795	48.875	48.955	49.035	49.116
650	49.116	49.196	49.276	49.356	49.436	49.517	49.597	49.677	49.757	49.837	49.917
660	49.917	49.997	50.077	50.157	50.238	50.318	50.398	50.478	50.558	50.638	50.718
670	50.718	50.798	50.878	50.958	51.038	51.118	51.197	51.277	51.357	51.437	51.517
680	51.517	51.597	51.677	51.757	51.837	51.916	51.996	52.076	52.156	52.236	52.315
690	52.315	52.395	52.475	52.555	52.634	52.714	52.794	52.873	52.953	53.033	53.112
700	53.112	53.192	53.272	53.351	53.431	53.51	53.59	53.67	53.749	53.829	53.908
710	53.908	53.988	54.067	54.147	54.226	54.306	54.385	54.465	54.544	54.624	54.703
720	54.703	54.782	54.862	54.941	55.021	55.1	55.179	55.259	55.338	55.417	55.497
730	55.497	55.576	55.655	55.734	55.814	55.893	55.972	56.051	56.131	56.21	56.289
740	56.289	56.368	56.447	56.526	56.606	56.685	56.764	56.843	56.922	57.001	57.08
750	57.08	57.159	57.238	57.317	57.396	57.475	57.554	57.633	57.712	57.791	57.87

Continues on next page

TABLE J.2 – cont.

°C	0	1	2	3	4	5	6	7	8	9	10
760	57.87	57.949	58.028	58.107	58.186	58.265	58.343	58.422	58.501	58.58	58.659
770	58.659	58.738	58.816	58.895	58.974	59.053	59.131	59.21	59.289	59.367	59.446
780	59.446	59.525	59.604	59.682	59.761	59.839	59.918	59.997	60.075	60.154	60.232
790	60.232	60.311	60.39	60.468	60.547	60.625	60.704	60.782	60.86	60.939	61.017
800	61.017	61.096	61.174	61.253	61.331	61.409	61.488	61.566	61.644	61.723	61.801
810	61.801	61.879	61.958	62.036	62.114	62.192	62.271	62.349	62.427	62.505	62.583
820	62.583	62.662	62.74	62.818	62.896	62.974	63.052	63.13	63.208	63.286	63.364
830	63.364	63.442	63.52	63.598	63.676	63.754	63.832	63.91	63.988	64.066	64.144
840	64.144	64.222	64.3	64.377	64.455	64.533	64.611	64.689	64.766	64.844	64.922
850	64.922	65	65.077	65.155	65.233	65.31	65.388	65.465	65.543	65.621	65.698
860	65.698	65.776	65.853	65.931	66.008	66.086	66.163	66.241	66.318	66.396	66.473
870	66.473	66.55	66.628	66.705	66.782	66.86	66.937	67.014	67.092	67.169	67.246
880	67.246	67.323	67.4	67.478	67.555	67.632	67.709	67.786	67.863	67.94	68.017
890	68.017	68.094	68.171	68.248	68.325	68.402	68.479	68.556	68.633	68.71	68.787
900	68.787	68.863	68.94	69.017	69.094	69.171	69.247	69.324	69.401	69.477	69.554
910	69.554	69.631	69.707	69.784	69.86	69.937	70.013	70.09	70.166	70.243	70.319
920	70.319	70.396	70.472	70.548	70.625	70.701	70.777	70.854	70.93	71.006	71.082
930	71.082	71.159	71.235	71.311	71.387	71.463	71.539	71.615	71.692	71.768	71.844
940	71.844	71.92	71.996	72.072	72.147	72.223	72.299	72.375	72.451	72.527	72.603
950	72.603	72.678	72.754	72.83	72.906	72.981	73.057	73.133	73.208	73.284	73.36
960	73.36	73.435	73.511	73.586	73.662	73.738	73.813	73.889	73.964	74.04	74.115
970	74.115	74.19	74.266	74.341	74.417	74.492	74.567	74.643	74.718	74.793	74.869
980	74.869	74.944	75.019	75.095	75.17	75.245	75.32	75.395	75.471	75.546	75.621
990	75.621	75.696	75.771	75.847	75.922	75.997	76.072	76.147	76.223	76.298	76.373
1000	76.373										

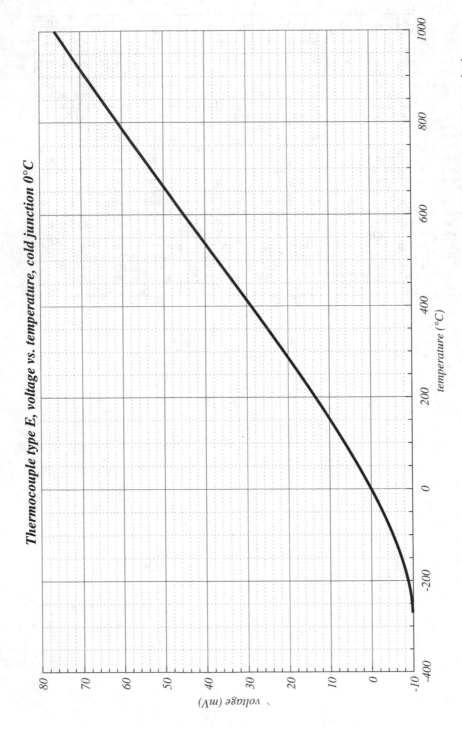

Thermocouple type E, voltage vs. temperature, cold junction 0°C

FIGURE J.1: Thermocouple type E, voltage vs. temperature, cold junction 0°C, based on data from [83]

K

Thermocouple type R

TABLE K.1: Thermocouple type R, negative temperatures, cold junction 0°C, data [83]

°C	0	-1	-2	-3	-4	-5	-6	-7	-8	-9	-10
-50	-0.226										
-40	-0.188	-0.192	-0.196	-0.2	-0.204	-0.208	-0.211	-0.215	-0.219	-0.223	-0.226
-30	-0.145	-0.15	-0.154	-0.158	-0.163	-0.167	-0.171	-0.175	-0.18	-0.184	-0.188
-20	-0.1	-0.105	-0.109	-0.114	-0.119	-0.123	-0.128	-0.132	-0.137	-0.141	-0.145
-10	-0.051	-0.056	-0.061	-0.066	-0.071	-0.076	-0.081	-0.086	-0.091	-0.095	-0.1
0	0	-0.005	-0.011	-0.016	-0.021	-0.026	-0.031	-0.036	-0.041	-0.046	-0.051

TABLE K.2: Thermocouple type R, positive temperatures, cold junction 0°C, data [83]

°C	0	1	2	3	4	5	6	7	8	9	10
0	0	0.005	0.011	0.016	0.021	0.027	0.032	0.038	0.043	0.049	0.054
10	0.054	0.06	0.065	0.071	0.077	0.082	0.088	0.094	0.1	0.105	0.111
20	0.111	0.117	0.123	0.129	0.135	0.141	0.147	0.153	0.159	0.165	0.171
30	0.171	0.177	0.183	0.189	0.195	0.201	0.207	0.214	0.22	0.226	0.232
40	0.232	0.239	0.245	0.251	0.258	0.264	0.271	0.277	0.284	0.29	0.296
50	0.296	0.303	0.31	0.316	0.323	0.329	0.336	0.343	0.349	0.356	0.363
60	0.363	0.369	0.376	0.383	0.39	0.397	0.403	0.41	0.417	0.424	0.431
70	0.431	0.438	0.445	0.452	0.459	0.466	0.473	0.48	0.487	0.494	0.501
80	0.501	0.508	0.516	0.523	0.53	0.537	0.544	0.552	0.559	0.566	0.573
90	0.573	0.581	0.588	0.595	0.603	0.61	0.618	0.625	0.632	0.64	0.647
100	0.647	0.655	0.662	0.67	0.677	0.685	0.693	0.7	0.708	0.715	0.723
110	0.723	0.731	0.738	0.746	0.754	0.761	0.769	0.777	0.785	0.792	0.8
120	0.8	0.808	0.816	0.824	0.832	0.839	0.847	0.855	0.863	0.871	0.879
130	0.879	0.887	0.895	0.903	0.911	0.919	0.927	0.935	0.943	0.951	0.959
140	0.959	0.967	0.976	0.984	0.992	1	1.008	1.016	1.025	1.033	1.041
150	1.041	1.049	1.058	1.066	1.074	1.082	1.091	1.099	1.107	1.116	1.124
160	1.124	1.132	1.141	1.149	1.158	1.166	1.175	1.183	1.191	1.2	1.208
170	1.208	1.217	1.225	1.234	1.242	1.251	1.26	1.268	1.277	1.285	1.294
180	1.294	1.303	1.311	1.32	1.329	1.337	1.346	1.355	1.363	1.372	1.381
190	1.381	1.389	1.398	1.407	1.416	1.425	1.433	1.442	1.451	1.46	1.469
200	1.469	1.477	1.486	1.495	1.504	1.513	1.522	1.531	1.54	1.549	1.558
210	1.558	1.567	1.575	1.584	1.593	1.602	1.611	1.62	1.629	1.639	1.648
220	1.648	1.657	1.666	1.675	1.684	1.693	1.702	1.711	1.72	1.729	1.739
230	1.739	1.748	1.757	1.766	1.775	1.784	1.794	1.803	1.812	1.821	1.831
240	1.831	1.84	1.849	1.858	1.868	1.877	1.886	1.895	1.905	1.914	1.923
250	1.923	1.933	1.942	1.951	1.961	1.97	1.98	1.989	1.998	2.008	2.017
260	2.017	2.027	2.036	2.046	2.055	2.064	2.074	2.083	2.093	2.102	2.112
270	2.112	2.121	2.131	2.14	2.15	2.159	2.169	2.179	2.188	2.198	2.207
280	2.207	2.217	2.226	2.236	2.246	2.255	2.265	2.275	2.284	2.294	2.304
290	2.304	2.313	2.323	2.333	2.342	2.352	2.362	2.371	2.381	2.391	2.401
300	2.401	2.41	2.42	2.43	2.44	2.449	2.459	2.469	2.479	2.488	2.498
310	2.498	2.508	2.518	2.528	2.538	2.547	2.557	2.567	2.577	2.587	2.597
320	2.597	2.607	2.617	2.626	2.636	2.646	2.656	2.666	2.676	2.686	2.696

Continues on next page

TABLE K.2 – cont.

°C	0	1	2	3	4	5	6	7	8	9	10
330	2.696	2.706	2.716	2.726	2.736	2.746	2.756	2.766	2.776	2.786	2.796
340	2.796	2.806	2.816	2.826	2.836	2.846	2.856	2.866	2.876	2.886	2.896
350	2.896	2.906	2.916	2.926	2.937	2.947	2.957	2.967	2.977	2.987	2.997
360	2.997	3.007	3.018	3.028	3.038	3.048	3.058	3.068	3.079	3.089	3.099
370	3.099	3.109	3.119	3.13	3.14	3.15	3.16	3.171	3.181	3.191	3.201
380	3.201	3.212	3.222	3.232	3.242	3.253	3.263	3.273	3.284	3.294	3.304
390	3.304	3.315	3.325	3.335	3.346	3.356	3.366	3.377	3.387	3.397	3.408
400	3.408	3.418	3.428	3.439	3.449	3.46	3.47	3.48	3.491	3.501	3.512
410	3.512	3.522	3.533	3.543	3.553	3.564	3.574	3.585	3.595	3.606	3.616
420	3.616	3.627	3.637	3.648	3.658	3.669	3.679	3.69	3.7	3.711	3.721
430	3.721	3.732	3.742	3.753	3.764	3.774	3.785	3.795	3.806	3.816	3.827
440	3.827	3.838	3.848	3.859	3.869	3.88	3.891	3.901	3.912	3.922	3.933
450	3.933	3.944	3.954	3.965	3.976	3.986	3.997	4.008	4.018	4.029	4.04
460	4.04	4.05	4.061	4.072	4.083	4.093	4.104	4.115	4.125	4.136	4.147
470	4.147	4.158	4.168	4.179	4.19	4.201	4.211	4.222	4.233	4.244	4.255
480	4.255	4.265	4.276	4.287	4.298	4.309	4.319	4.33	4.341	4.352	4.363
490	4.363	4.373	4.384	4.395	4.406	4.417	4.428	4.439	4.449	4.46	4.471
500	4.471	4.482	4.493	4.504	4.515	4.526	4.537	4.548	4.558	4.569	4.58
510	4.58	4.591	4.602	4.613	4.624	4.635	4.646	4.657	4.668	4.679	4.69
520	4.69	4.701	4.712	4.723	4.734	4.745	4.756	4.767	4.778	4.789	4.8
530	4.8	4.811	4.822	4.833	4.844	4.855	4.866	4.877	4.888	4.899	4.91
540	4.91	4.922	4.933	4.944	4.955	4.966	4.977	4.988	4.999	5.01	5.021
550	5.021	5.033	5.044	5.055	5.066	5.077	5.088	5.099	5.111	5.122	5.133
560	5.133	5.144	5.155	5.166	5.178	5.189	5.2	5.211	5.222	5.234	5.245
570	5.245	5.256	5.267	5.279	5.29	5.301	5.312	5.323	5.335	5.346	5.357
580	5.357	5.369	5.38	5.391	5.402	5.414	5.425	5.436	5.448	5.459	5.47
590	5.47	5.481	5.493	5.504	5.515	5.527	5.538	5.549	5.561	5.572	5.583
600	5.583	5.595	5.606	5.618	5.629	5.64	5.652	5.663	5.674	5.686	5.697
610	5.697	5.709	5.72	5.731	5.743	5.754	5.766	5.777	5.789	5.8	5.812
620	5.812	5.823	5.834	5.846	5.857	5.869	5.88	5.892	5.903	5.915	5.926
630	5.926	5.938	5.949	5.961	5.972	5.984	5.995	6.007	6.018	6.03	6.041
640	6.041	6.053	6.065	6.076	6.088	6.099	6.111	6.122	6.134	6.146	6.157
650	6.157	6.169	6.18	6.192	6.204	6.215	6.227	6.238	6.25	6.262	6.273
660	6.273	6.285	6.297	6.308	6.32	6.332	6.343	6.355	6.367	6.378	6.39
670	6.39	6.402	6.413	6.425	6.437	6.448	6.46	6.472	6.484	6.495	6.507
680	6.507	6.519	6.531	6.542	6.554	6.566	6.578	6.589	6.601	6.613	6.625
690	6.625	6.636	6.648	6.66	6.672	6.684	6.695	6.707	6.719	6.731	6.743
700	6.743	6.755	6.766	6.778	6.79	6.802	6.814	6.826	6.838	6.849	6.861
710	6.861	6.873	6.885	6.897	6.909	6.921	6.933	6.945	6.956	6.968	6.98
720	6.98	6.992	7.004	7.016	7.028	7.04	7.052	7.064	7.076	7.088	7.1
730	7.1	7.112	7.124	7.136	7.148	7.16	7.172	7.184	7.196	7.208	7.22
740	7.22	7.232	7.244	7.256	7.268	7.28	7.292	7.304	7.316	7.328	7.34
750	7.34	7.352	7.364	7.376	7.389	7.401	7.413	7.425	7.437	7.449	7.461
760	7.461	7.473	7.485	7.498	7.51	7.522	7.534	7.546	7.558	7.57	7.583
770	7.583	7.595	7.607	7.619	7.631	7.644	7.656	7.668	7.68	7.692	7.705
780	7.705	7.717	7.729	7.741	7.753	7.766	7.778	7.79	7.802	7.815	7.827
790	7.827	7.839	7.851	7.864	7.876	7.888	7.901	7.913	7.925	7.938	7.95
800	7.95	7.962	7.974	7.987	7.999	8.011	8.024	8.036	8.048	8.061	8.073
810	8.073	8.086	8.098	8.11	8.123	8.135	8.147	8.16	8.172	8.185	8.197
820	8.197	8.209	8.222	8.234	8.247	8.259	8.272	8.284	8.296	8.309	8.321
830	8.321	8.334	8.346	8.359	8.371	8.384	8.396	8.409	8.421	8.434	8.446
840	8.446	8.459	8.471	8.484	8.496	8.509	8.521	8.534	8.546	8.559	8.571
850	8.571	8.584	8.597	8.609	8.622	8.634	8.647	8.659	8.672	8.685	8.697
860	8.697	8.71	8.722	8.735	8.748	8.76	8.773	8.785	8.798	8.811	8.823
870	8.823	8.836	8.849	8.861	8.874	8.887	8.899	8.912	8.925	8.937	8.95
880	8.95	8.963	8.975	8.988	9.001	9.014	9.026	9.039	9.052	9.065	9.077
890	9.077	9.09	9.103	9.115	9.128	9.141	9.154	9.167	9.179	9.192	9.205
900	9.205	9.218	9.23	9.243	9.256	9.269	9.282	9.294	9.307	9.32	9.333
910	9.333	9.346	9.359	9.371	9.384	9.397	9.41	9.423	9.436	9.449	9.461
920	9.461	9.474	9.487	9.5	9.513	9.526	9.539	9.552	9.565	9.578	9.59
930	9.59	9.603	9.616	9.629	9.642	9.655	9.668	9.681	9.694	9.707	9.72
940	9.72	9.733	9.746	9.759	9.772	9.785	9.798	9.811	9.824	9.837	9.85
950	9.85	9.863	9.876	9.889	9.902	9.915	9.928	9.941	9.954	9.967	9.98
960	9.98	9.993	10.006	10.019	10.032	10.046	10.059	10.072	10.085	10.098	10.111
970	10.111	10.124	10.137	10.15	10.163	10.177	10.19	10.203	10.216	10.229	10.242

Continues on next page

TABLE K.2 – cont.

°C	0	1	2	3	4	5	6	7	8	9	10
980	10.242	10.255	10.268	10.282	10.295	10.308	10.321	10.334	10.347	10.361	10.374
990	10.374	10.387	10.4	10.413	10.427	10.44	10.453	10.466	10.48	10.493	10.506
1000	10.506	10.519	10.532	10.546	10.559	10.572	10.585	10.599	10.612	10.625	10.638
1010	10.638	10.652	10.665	10.678	10.692	10.705	10.718	10.731	10.745	10.758	10.771
1020	10.771	10.785	10.798	10.811	10.825	10.838	10.851	10.865	10.878	10.891	10.905
1030	10.905	10.918	10.932	10.945	10.958	10.972	10.985	10.998	11.012	11.025	11.039
1040	11.039	11.052	11.065	11.079	11.092	11.106	11.119	11.132	11.146	11.159	11.173
1050	11.173	11.186	11.2	11.213	11.227	11.24	11.253	11.267	11.28	11.294	11.307
1060	11.307	11.321	11.334	11.348	11.361	11.375	11.388	11.402	11.415	11.429	11.442
1070	11.442	11.456	11.469	11.483	11.496	11.51	11.524	11.537	11.551	11.564	11.578
1080	11.578	11.591	11.605	11.618	11.632	11.646	11.659	11.673	11.686	11.7	11.714
1090	11.714	11.727	11.741	11.754	11.768	11.782	11.795	11.809	11.822	11.836	11.85
1100	11.85	11.863	11.877	11.891	11.904	11.918	11.931	11.945	11.959	11.972	11.986
1110	11.986	12	12.013	12.027	12.041	12.054	12.068	12.082	12.096	12.109	12.123
1120	12.123	12.137	12.15	12.164	12.178	12.191	12.205	12.219	12.233	12.246	12.26
1130	12.26	12.274	12.288	12.301	12.315	12.329	12.342	12.356	12.37	12.384	12.397
1140	12.397	12.411	12.425	12.439	12.453	12.466	12.48	12.494	12.508	12.521	12.535
1150	12.535	12.549	12.563	12.577	12.59	12.604	12.618	12.632	12.646	12.659	12.673
1160	12.673	12.687	12.701	12.715	12.729	12.742	12.756	12.77	12.784	12.798	12.812
1170	12.812	12.825	12.839	12.853	12.867	12.881	12.895	12.909	12.922	12.936	12.95
1180	12.95	12.964	12.978	12.992	13.006	13.019	13.033	13.047	13.061	13.075	13.089
1190	13.089	13.103	13.117	13.131	13.145	13.158	13.172	13.186	13.2	13.214	13.228
1200	13.228	13.242	13.256	13.27	13.284	13.298	13.311	13.325	13.339	13.353	13.367
1210	13.367	13.381	13.395	13.409	13.423	13.437	13.451	13.465	13.479	13.493	13.507
1220	13.507	13.521	13.535	13.549	13.563	13.577	13.59	13.604	13.618	13.632	13.646
1230	13.646	13.66	13.674	13.688	13.702	13.716	13.73	13.744	13.758	13.772	13.786
1240	13.786	13.8	13.814	13.828	13.842	13.856	13.87	13.884	13.898	13.912	13.926
1250	13.926	13.94	13.954	13.968	13.982	13.996	14.01	14.024	14.038	14.052	14.066
1260	14.066	14.081	14.095	14.109	14.123	14.137	14.151	14.165	14.179	14.193	14.207
1270	14.207	14.221	14.235	14.249	14.263	14.277	14.291	14.305	14.319	14.333	14.347
1280	14.347	14.361	14.375	14.39	14.404	14.418	14.432	14.446	14.46	14.474	14.488
1290	14.488	14.502	14.516	14.53	14.544	14.558	14.572	14.586	14.601	14.615	14.629
1300	14.629	14.643	14.657	14.671	14.685	14.699	14.713	14.727	14.741	14.755	14.77
1310	14.77	14.784	14.798	14.812	14.826	14.84	14.854	14.868	14.882	14.896	14.911
1320	14.911	14.925	14.939	14.953	14.967	14.981	14.995	15.009	15.023	15.037	15.052
1330	15.052	15.066	15.08	15.094	15.108	15.122	15.136	15.15	15.164	15.179	15.193
1340	15.193	15.207	15.221	15.235	15.249	15.263	15.277	15.291	15.306	15.32	15.334
1350	15.334	15.348	15.362	15.376	15.39	15.404	15.419	15.433	15.447	15.461	15.475
1360	15.475	15.489	15.503	15.517	15.531	15.546	15.56	15.574	15.588	15.602	15.616
1370	15.616	15.63	15.645	15.659	15.673	15.687	15.701	15.715	15.729	15.743	15.758
1380	15.758	15.772	15.786	15.8	15.814	15.828	15.842	15.856	15.871	15.885	15.899
1390	15.899	15.913	15.927	15.941	15.955	15.969	15.984	15.998	16.012	16.026	16.04
1400	16.04	16.054	16.068	16.082	16.097	16.111	16.125	16.139	16.153	16.167	16.181
1410	16.181	16.196	16.21	16.224	16.238	16.252	16.266	16.28	16.294	16.309	16.323
1420	16.323	16.337	16.351	16.365	16.379	16.393	16.407	16.422	16.436	16.45	16.464
1430	16.464	16.478	16.492	16.506	16.52	16.534	16.549	16.563	16.577	16.591	16.605
1440	16.605	16.619	16.633	16.647	16.662	16.676	16.69	16.704	16.718	16.732	16.746
1450	16.746	16.76	16.774	16.789	16.803	16.817	16.831	16.845	16.859	16.873	16.887
1460	16.887	16.901	16.915	16.93	16.944	16.958	16.972	16.986	17	17.014	17.028
1470	17.028	17.042	17.056	17.071	17.085	17.099	17.113	17.127	17.141	17.155	17.169
1480	17.169	17.183	17.197	17.211	17.225	17.24	17.254	17.268	17.282	17.296	17.31
1490	17.31	17.324	17.338	17.352	17.366	17.38	17.394	17.408	17.423	17.437	17.451
1500	17.451	17.465	17.479	17.493	17.507	17.521	17.535	17.549	17.563	17.577	17.591
1510	17.591	17.605	17.619	17.633	17.647	17.661	17.676	17.69	17.704	17.718	17.732
1520	17.732	17.746	17.76	17.774	17.788	17.802	17.816	17.83	17.844	17.858	17.872
1530	17.872	17.886	17.9	17.914	17.928	17.942	17.956	17.97	17.984	17.998	18.012
1540	18.012	18.026	18.04	18.054	18.068	18.082	18.096	18.11	18.124	18.138	18.152
1550	18.152	18.166	18.18	18.194	18.208	18.222	18.236	18.25	18.264	18.278	18.292
1560	18.292	18.306	18.32	18.334	18.348	18.362	18.376	18.39	18.404	18.417	18.431
1570	18.431	18.445	18.459	18.473	18.487	18.501	18.515	18.529	18.543	18.557	18.571
1580	18.571	18.585	18.599	18.613	18.627	18.64	18.654	18.668	18.682	18.696	18.71
1590	18.71	18.724	18.738	18.752	18.766	18.779	18.793	18.807	18.821	18.835	18.849
1600	18.849	18.863	18.877	18.891	18.904	18.918	18.932	18.946	18.96	18.974	18.988
1610	18.988	19.002	19.015	19.029	19.043	19.057	19.071	19.085	19.098	19.112	19.126
1620	19.126	19.14	19.154	19.168	19.181	19.195	19.209	19.223	19.237	19.25	19.264

Continues on next page

TABLE K.2 – cont.

°C	0	1	2	3	4	5	6	7	8	9	10
1630	19.264	19.278	19.292	19.306	19.319	19.333	19.347	19.361	19.375	19.388	19.402
1640	19.402	19.416	19.43	19.444	19.457	19.471	19.485	19.499	19.512	19.526	19.54
1650	19.54	19.554	19.567	19.581	19.595	19.609	19.622	19.636	19.65	19.663	19.677
1660	19.677	19.691	19.705	19.718	19.732	19.746	19.759	19.773	19.787	19.8	19.814
1670	19.814	19.828	19.841	19.855	19.869	19.882	19.896	19.91	19.923	19.937	19.951
1680	19.951	19.964	19.978	19.992	20.005	20.019	20.032	20.046	20.06	20.073	20.087
1690	20.087	20.1	20.114	20.127	20.141	20.154	20.168	20.181	20.195	20.208	20.222
1700	20.222	20.235	20.249	20.262	20.275	20.289	20.302	20.316	20.329	20.342	20.356
1710	20.356	20.369	20.382	20.396	20.409	20.422	20.436	20.449	20.462	20.475	20.488
1720	20.488	20.502	20.515	20.528	20.541	20.554	20.567	20.581	20.594	20.607	20.62
1730	20.62	20.633	20.646	20.659	20.672	20.685	20.698	20.711	20.724	20.736	20.749
1740	20.749	20.762	20.775	20.788	20.801	20.813	20.826	20.839	20.852	20.864	20.877
1750	20.877	20.89	20.902	20.915	20.928	20.94	20.953	20.965	20.978	20.99	21.003
1760	21.003	21.015	21.027	21.04	21.052	21.065	21.077	21.089	21.101		

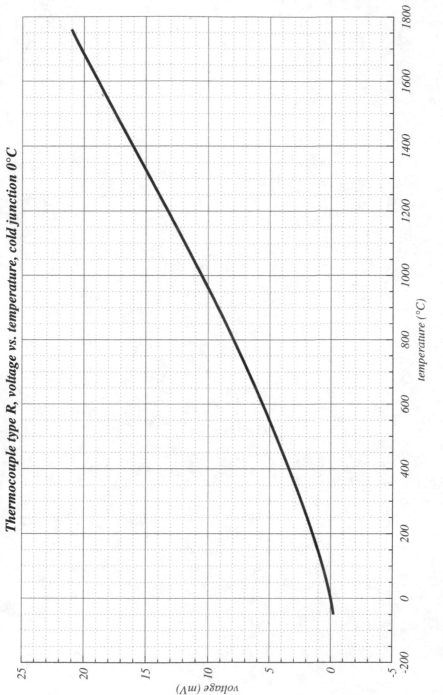

FIGURE K.1: Thermocouple type R, voltage vs. temperature, cold junction 0°C, based on data from [83]

L

Thermocouple type S

TABLE L.1: Thermocouple type S, negative temperatures, cold junction 0°C, data [83]

°C	0	-1	-2	-3	-4	-5	-6	-7	-8	-9	-10
-50	-0.236										
-40	-0.194	-0.199	-0.203	-0.207	-0.211	-0.215	-0.219	-0.224	-0.228	-0.232	-0.236
-30	-0.15	-0.155	-0.159	-0.164	-0.168	-0.173	-0.177	-0.181	-0.186	-0.19	-0.194
-20	-0.103	-0.108	-0.113	-0.117	-0.122	-0.127	-0.132	-0.136	-0.141	-0.146	-0.15
-10	-0.053	-0.058	-0.063	-0.068	-0.073	-0.078	-0.083	-0.088	-0.093	-0.098	-0.103
0	0	-0.005	-0.011	-0.016	-0.021	-0.027	-0.032	-0.037	-0.042	-0.048	-0.053

TABLE L.2: Thermocouple type S, positive temperatures, cold junction 0°C, data [83]

°C	0	1	2	3	4	5	6	7	8	9	10
0	0	0.005	0.011	0.016	0.022	0.027	0.033	0.038	0.044	0.05	0.055
10	0.055	0.061	0.067	0.072	0.078	0.084	0.09	0.095	0.101	0.107	0.113
20	0.113	0.119	0.125	0.131	0.137	0.143	0.149	0.155	0.161	0.167	0.173
30	0.173	0.179	0.185	0.191	0.197	0.204	0.21	0.216	0.222	0.229	0.235
40	0.235	0.241	0.248	0.254	0.26	0.267	0.273	0.28	0.286	0.292	0.299
50	0.299	0.305	0.312	0.319	0.325	0.332	0.338	0.345	0.352	0.358	0.365
60	0.365	0.372	0.378	0.385	0.392	0.399	0.405	0.412	0.419	0.426	0.433
70	0.433	0.44	0.446	0.453	0.46	0.467	0.474	0.481	0.488	0.495	0.502
80	0.502	0.509	0.516	0.523	0.53	0.538	0.545	0.552	0.559	0.566	0.573
90	0.573	0.58	0.588	0.595	0.602	0.609	0.617	0.624	0.631	0.639	0.646
100	0.646	0.653	0.661	0.668	0.675	0.683	0.69	0.698	0.705	0.713	0.72
110	0.72	0.727	0.735	0.743	0.75	0.758	0.765	0.773	0.78	0.788	0.795
120	0.795	0.803	0.811	0.818	0.826	0.834	0.841	0.849	0.857	0.865	0.872
130	0.872	0.88	0.888	0.896	0.903	0.911	0.919	0.927	0.935	0.942	0.95
140	0.95	0.958	0.966	0.974	0.982	0.99	0.998	1.006	1.013	1.021	1.029
150	1.029	1.037	1.045	1.053	1.061	1.069	1.077	1.085	1.094	1.102	1.11
160	1.11	1.118	1.126	1.134	1.142	1.15	1.158	1.167	1.175	1.183	1.191
170	1.191	1.199	1.207	1.216	1.224	1.232	1.24	1.249	1.257	1.265	1.273
180	1.273	1.282	1.29	1.298	1.307	1.315	1.323	1.332	1.34	1.348	1.357
190	1.357	1.365	1.373	1.382	1.39	1.399	1.407	1.415	1.424	1.432	1.441
200	1.441	1.449	1.458	1.466	1.475	1.483	1.492	1.5	1.509	1.517	1.526
210	1.526	1.534	1.543	1.551	1.56	1.569	1.577	1.586	1.594	1.603	1.612
220	1.612	1.62	1.629	1.638	1.646	1.655	1.663	1.672	1.681	1.69	1.698
230	1.698	1.707	1.716	1.724	1.733	1.742	1.751	1.759	1.768	1.777	1.786
240	1.786	1.794	1.803	1.812	1.821	1.829	1.838	1.847	1.856	1.865	1.874
250	1.874	1.882	1.891	1.9	1.909	1.918	1.927	1.936	1.944	1.953	1.962
260	1.962	1.971	1.98	1.989	1.998	2.007	2.016	2.025	2.034	2.043	2.052
270	2.052	2.061	2.07	2.078	2.087	2.096	2.105	2.114	2.123	2.132	2.141
280	2.141	2.151	2.16	2.169	2.178	2.187	2.196	2.205	2.214	2.223	2.232
290	2.232	2.241	2.25	2.259	2.268	2.277	2.287	2.296	2.305	2.314	2.323
300	2.323	2.332	2.341	2.35	2.36	2.369	2.378	2.387	2.396	2.405	2.415
310	2.415	2.424	2.433	2.442	2.451	2.461	2.47	2.479	2.488	2.497	2.507
320	2.507	2.516	2.525	2.534	2.544	2.553	2.562	2.571	2.581	2.59	2.599

Continues on next page

TABLE L.2 – cont.

°C	0	1	2	3	4	5	6	7	8	9	10
330	2.599	2.609	2.618	2.627	2.636	2.646	2.655	2.664	2.674	2.683	2.692
340	2.692	2.702	2.711	2.72	2.73	2.739	2.748	2.758	2.767	2.776	2.786
350	2.786	2.795	2.805	2.814	2.823	2.833	2.842	2.851	2.861	2.87	2.88
360	2.88	2.889	2.899	2.908	2.917	2.927	2.936	2.946	2.955	2.965	2.974
370	2.974	2.983	2.993	3.002	3.012	3.021	3.031	3.04	3.05	3.059	3.069
380	3.069	3.078	3.088	3.097	3.107	3.116	3.126	3.135	3.145	3.154	3.164
390	3.164	3.173	3.183	3.192	3.202	3.212	3.221	3.231	3.24	3.25	3.259
400	3.259	3.269	3.279	3.288	3.298	3.307	3.317	3.326	3.336	3.346	3.355
410	3.355	3.365	3.374	3.384	3.394	3.403	3.413	3.423	3.432	3.442	3.451
420	3.451	3.461	3.471	3.48	3.49	3.5	3.509	3.519	3.529	3.538	3.548
430	3.548	3.558	3.567	3.577	3.587	3.596	3.606	3.616	3.626	3.635	3.645
440	3.645	3.655	3.664	3.674	3.684	3.694	3.703	3.713	3.723	3.732	3.742
450	3.742	3.752	3.762	3.771	3.781	3.791	3.801	3.81	3.82	3.83	3.84
460	3.84	3.85	3.859	3.869	3.879	3.889	3.898	3.908	3.918	3.928	3.938
470	3.938	3.947	3.957	3.967	3.977	3.987	3.997	4.006	4.016	4.026	4.036
480	4.036	4.046	4.056	4.065	4.075	4.085	4.095	4.105	4.115	4.125	4.134
490	4.134	4.144	4.154	4.164	4.174	4.184	4.194	4.204	4.213	4.223	4.233
500	4.233	4.243	4.253	4.263	4.273	4.283	4.293	4.303	4.313	4.323	4.332
510	4.332	4.342	4.352	4.362	4.372	4.382	4.392	4.402	4.412	4.422	4.432
520	4.432	4.442	4.452	4.462	4.472	4.482	4.492	4.502	4.512	4.522	4.532
530	4.532	4.542	4.552	4.562	4.572	4.582	4.592	4.602	4.612	4.622	4.632
540	4.632	4.642	4.652	4.662	4.672	4.682	4.692	4.702	4.712	4.722	4.732
550	4.732	4.742	4.752	4.762	4.772	4.782	4.793	4.803	4.813	4.823	4.833
560	4.833	4.843	4.853	4.863	4.873	4.883	4.893	4.904	4.914	4.924	4.934
570	4.934	4.944	4.954	4.964	4.974	4.984	4.995	5.005	5.015	5.025	5.035
580	5.035	5.045	5.055	5.066	5.076	5.086	5.096	5.106	5.116	5.127	5.137
590	5.137	5.147	5.157	5.167	5.178	5.188	5.198	5.208	5.218	5.228	5.239
600	5.239	5.249	5.259	5.269	5.28	5.29	5.3	5.31	5.32	5.331	5.341
610	5.341	5.351	5.361	5.372	5.382	5.392	5.402	5.413	5.423	5.433	5.443
620	5.443	5.454	5.464	5.474	5.485	5.495	5.505	5.515	5.526	5.536	5.546
630	5.546	5.557	5.567	5.577	5.588	5.598	5.608	5.618	5.629	5.639	5.649
640	5.649	5.66	5.67	5.68	5.691	5.701	5.712	5.722	5.732	5.743	5.753
650	5.753	5.763	5.774	5.784	5.794	5.805	5.815	5.826	5.836	5.846	5.857
660	5.857	5.867	5.878	5.888	5.898	5.909	5.919	5.93	5.94	5.95	5.961
670	5.961	5.971	5.982	5.992	6.003	6.013	6.024	6.034	6.044	6.055	6.065
680	6.065	6.076	6.086	6.097	6.107	6.118	6.128	6.139	6.149	6.16	6.17
690	6.17	6.181	6.191	6.202	6.212	6.223	6.233	6.244	6.254	6.265	6.275
700	6.275	6.286	6.296	6.307	6.317	6.328	6.338	6.349	6.36	6.37	6.381
710	6.381	6.391	6.402	6.412	6.423	6.434	6.444	6.455	6.465	6.476	6.486
720	6.486	6.497	6.508	6.518	6.529	6.539	6.55	6.561	6.571	6.582	6.593
730	6.593	6.603	6.614	6.624	6.635	6.646	6.656	6.667	6.678	6.688	6.699
740	6.699	6.71	6.72	6.731	6.742	6.752	6.763	6.774	6.784	6.795	6.806
750	6.806	6.817	6.827	6.838	6.849	6.859	6.87	6.881	6.892	6.902	6.913
760	6.913	6.924	6.934	6.945	6.956	6.967	6.977	6.988	6.999	7.01	7.02
770	7.02	7.031	7.042	7.053	7.064	7.074	7.085	7.096	7.107	7.117	7.128
780	7.128	7.139	7.15	7.161	7.172	7.182	7.193	7.204	7.215	7.226	7.236
790	7.236	7.247	7.258	7.269	7.28	7.291	7.302	7.312	7.323	7.334	7.345
800	7.345	7.356	7.367	7.378	7.388	7.399	7.41	7.421	7.432	7.443	7.454
810	7.454	7.465	7.476	7.487	7.497	7.508	7.519	7.53	7.541	7.552	7.563
820	7.563	7.574	7.585	7.596	7.607	7.618	7.629	7.64	7.651	7.662	7.673
830	7.673	7.684	7.695	7.706	7.717	7.728	7.739	7.75	7.761	7.772	7.783
840	7.783	7.794	7.805	7.816	7.827	7.838	7.849	7.86	7.871	7.882	7.893
850	7.893	7.904	7.915	7.926	7.937	7.948	7.959	7.97	7.981	7.992	8.003
860	8.003	8.014	8.026	8.037	8.048	8.059	8.07	8.081	8.092	8.103	8.114
870	8.114	8.125	8.137	8.148	8.159	8.17	8.181	8.192	8.203	8.214	8.226
880	8.226	8.237	8.248	8.259	8.27	8.281	8.293	8.304	8.315	8.326	8.337
890	8.337	8.348	8.36	8.371	8.382	8.393	8.404	8.416	8.427	8.438	8.449
900	8.449	8.46	8.472	8.483	8.494	8.505	8.517	8.528	8.539	8.55	8.562
910	8.562	8.573	8.584	8.595	8.607	8.618	8.629	8.64	8.652	8.663	8.674
920	8.674	8.685	8.697	8.708	8.719	8.731	8.742	8.753	8.765	8.776	8.787
930	8.787	8.798	8.81	8.821	8.832	8.844	8.855	8.866	8.878	8.889	8.9
940	8.9	8.912	8.923	8.935	8.946	8.957	8.969	8.98	8.991	9.003	9.014
950	9.014	9.025	9.037	9.048	9.06	9.071	9.082	9.094	9.105	9.117	9.128
960	9.128	9.139	9.151	9.162	9.174	9.185	9.197	9.208	9.219	9.231	9.242
970	9.242	9.254	9.265	9.277	9.288	9.3	9.311	9.323	9.334	9.345	9.357

Continues on next page

TABLE L.2 – cont.

°C	0	1	2	3	4	5	6	7	8	9	10
980	9.357	9.368	9.38	9.391	9.403	9.414	9.426	9.437	9.449	9.46	9.472
990	9.472	9.483	9.495	9.506	9.518	9.529	9.541	9.552	9.564	9.576	9.587
1000	9.587	9.599	9.61	9.622	9.633	9.645	9.656	9.668	9.68	9.691	9.703
1010	9.703	9.714	9.726	9.737	9.749	9.761	9.772	9.784	9.795	9.807	9.819
1020	9.819	9.83	9.842	9.853	9.865	9.877	9.888	9.9	9.911	9.923	9.935
1030	9.935	9.946	9.958	9.97	9.981	9.993	10.005	10.016	10.028	10.04	10.051
1040	10.051	10.063	10.075	10.086	10.098	10.11	10.121	10.133	10.145	10.156	10.168
1050	10.168	10.18	10.191	10.203	10.215	10.227	10.238	10.25	10.262	10.273	10.285
1060	10.285	10.297	10.309	10.32	10.332	10.344	10.356	10.367	10.379	10.391	10.403
1070	10.403	10.414	10.426	10.438	10.45	10.461	10.473	10.485	10.497	10.509	10.52
1080	10.52	10.532	10.544	10.556	10.567	10.579	10.591	10.603	10.615	10.626	10.638
1090	10.638	10.65	10.662	10.674	10.686	10.697	10.709	10.721	10.733	10.745	10.757
1100	10.757	10.768	10.78	10.792	10.804	10.816	10.828	10.839	10.851	10.863	10.875
1110	10.875	10.887	10.899	10.911	10.922	10.934	10.946	10.958	10.97	10.982	10.994
1120	10.994	11.006	11.017	11.029	11.041	11.053	11.065	11.077	11.089	11.101	11.113
1130	11.113	11.125	11.136	11.148	11.16	11.172	11.184	11.196	11.208	11.22	11.232
1140	11.232	11.244	11.256	11.268	11.28	11.291	11.303	11.315	11.327	11.339	11.351
1150	11.351	11.363	11.375	11.387	11.399	11.411	11.423	11.435	11.447	11.459	11.471
1160	11.471	11.483	11.495	11.507	11.519	11.531	11.542	11.554	11.566	11.578	11.59
1170	11.59	11.602	11.614	11.626	11.638	11.65	11.662	11.674	11.686	11.698	11.71
1180	11.71	11.722	11.734	11.746	11.758	11.77	11.782	11.794	11.806	11.818	11.83
1190	11.83	11.842	11.854	11.866	11.878	11.89	11.902	11.914	11.926	11.939	11.951
1200	11.951	11.963	11.975	11.987	11.999	12.011	12.023	12.035	12.047	12.059	12.071
1210	12.071	12.083	12.095	12.107	12.119	12.131	12.143	12.155	12.167	12.179	12.191
1220	12.191	12.203	12.216	12.228	12.24	12.252	12.264	12.276	12.288	12.3	12.312
1230	12.312	12.324	12.336	12.348	12.36	12.372	12.384	12.397	12.409	12.421	12.433
1240	12.433	12.445	12.457	12.469	12.481	12.493	12.505	12.517	12.529	12.542	12.554
1250	12.554	12.566	12.578	12.59	12.602	12.614	12.626	12.638	12.65	12.662	12.675
1260	12.675	12.687	12.699	12.711	12.723	12.735	12.747	12.759	12.771	12.783	12.796
1270	12.796	12.808	12.82	12.832	12.844	12.856	12.868	12.88	12.892	12.905	12.917
1280	12.917	12.929	12.941	12.953	12.965	12.977	12.989	13.001	13.014	13.026	13.038
1290	13.038	13.05	13.062	13.074	13.086	13.098	13.111	13.123	13.135	13.147	13.159
1300	13.159	13.171	13.183	13.195	13.208	13.22	13.232	13.244	13.256	13.268	13.28
1310	13.28	13.292	13.305	13.317	13.329	13.341	13.353	13.365	13.377	13.39	13.402
1320	13.402	13.414	13.426	13.438	13.45	13.462	13.474	13.487	13.499	13.511	13.523
1330	13.523	13.535	13.547	13.559	13.572	13.584	13.596	13.608	13.62	13.632	13.644
1340	13.644	13.657	13.669	13.681	13.693	13.705	13.717	13.729	13.742	13.754	13.766
1350	13.766	13.778	13.79	13.802	13.814	13.826	13.839	13.851	13.863	13.875	13.887
1360	13.887	13.899	13.911	13.924	13.936	13.948	13.96	13.972	13.984	13.996	14.009
1370	14.009	14.021	14.033	14.045	14.057	14.069	14.081	14.094	14.106	14.118	14.13
1380	14.13	14.142	14.154	14.166	14.178	14.191	14.203	14.215	14.227	14.239	14.251
1390	14.251	14.263	14.276	14.288	14.3	14.312	14.324	14.336	14.348	14.36	14.373
1400	14.373	14.385	14.397	14.409	14.421	14.433	14.445	14.457	14.47	14.482	14.494
1410	14.494	14.506	14.518	14.53	14.542	14.554	14.567	14.579	14.591	14.603	14.615
1420	14.615	14.627	14.639	14.651	14.664	14.676	14.688	14.7	14.712	14.724	14.736
1430	14.736	14.748	14.76	14.773	14.785	14.797	14.809	14.821	14.833	14.845	14.857
1440	14.857	14.869	14.881	14.894	14.906	14.918	14.93	14.942	14.954	14.966	14.978
1450	14.978	14.99	15.002	15.015	15.027	15.039	15.051	15.063	15.075	15.087	15.099
1460	15.099	15.111	15.123	15.135	15.148	15.16	15.172	15.184	15.196	15.208	15.22
1470	15.22	15.232	15.244	15.256	15.268	15.28	15.292	15.304	15.317	15.329	15.341
1480	15.341	15.353	15.365	15.377	15.389	15.401	15.413	15.425	15.437	15.449	15.461
1490	15.461	15.473	15.485	15.497	15.509	15.521	15.534	15.546	15.558	15.57	15.582
1500	15.582	15.594	15.606	15.618	15.63	15.642	15.654	15.666	15.678	15.69	15.702
1510	15.702	15.714	15.726	15.738	15.75	15.762	15.774	15.786	15.798	15.81	15.822
1520	15.822	15.834	15.846	15.858	15.87	15.882	15.894	15.906	15.918	15.93	15.942
1530	15.942	15.954	15.966	15.978	15.99	16.002	16.014	16.026	16.038	16.05	16.062
1540	16.062	16.074	16.086	16.098	16.11	16.122	16.134	16.146	16.158	16.17	16.182
1550	16.182	16.194	16.205	16.217	16.229	16.241	16.253	16.265	16.277	16.289	16.301
1560	16.301	16.313	16.325	16.337	16.349	16.361	16.373	16.385	16.396	16.408	16.42
1570	16.42	16.432	16.444	16.456	16.468	16.48	16.492	16.504	16.516	16.527	16.539
1580	16.539	16.551	16.563	16.575	16.587	16.599	16.611	16.623	16.634	16.646	16.658
1590	16.658	16.67	16.682	16.694	16.706	16.718	16.729	16.741	16.753	16.765	16.777
1600	16.777	16.789	16.801	16.812	16.824	16.836	16.848	16.86	16.872	16.883	16.895
1610	16.895	16.907	16.919	16.931	16.943	16.954	16.966	16.978	16.99	17.002	17.013
1620	17.013	17.025	17.037	17.049	17.061	17.072	17.084	17.096	17.108	17.12	17.131

Continues on next page

TABLE L.2 – cont.

°C	0	1	2	3	4	5	6	7	8	9	10
1630	17.131	17.143	17.155	17.167	17.178	17.19	17.202	17.214	17.225	17.237	17.249
1640	17.249	17.261	17.272	17.284	17.296	17.308	17.319	17.331	17.343	17.355	17.366
1650	17.366	17.378	17.39	17.401	17.413	17.425	17.437	17.448	17.46	17.472	17.483
1660	17.483	17.495	17.507	17.518	17.53	17.542	17.553	17.565	17.577	17.588	17.6
1670	17.6	17.612	17.623	17.635	17.647	17.658	17.67	17.682	17.693	17.705	17.717
1680	17.717	17.728	17.74	17.751	17.763	17.775	17.786	17.798	17.809	17.821	17.832
1690	17.832	17.844	17.855	17.867	17.878	17.89	17.901	17.913	17.924	17.936	17.947
1700	17.947	17.959	17.97	17.982	17.993	18.004	18.016	18.027	18.039	18.05	18.061
1710	18.061	18.073	18.084	18.095	18.107	18.118	18.129	18.14	18.152	18.163	18.174
1720	18.174	18.185	18.196	18.208	18.219	18.23	18.241	18.252	18.263	18.274	18.285
1730	18.285	18.297	18.308	18.319	18.33	18.341	18.352	18.362	18.373	18.384	18.395
1740	18.395	18.406	18.417	18.428	18.439	18.449	18.46	18.471	18.482	18.493	18.503
1750	18.503	18.514	18.525	18.535	18.546	18.557	18.567	18.578	18.588	18.599	18.609
1760	18.609	18.62	18.63	18.641	18.651	18.661	18.672	18.682	18.693		

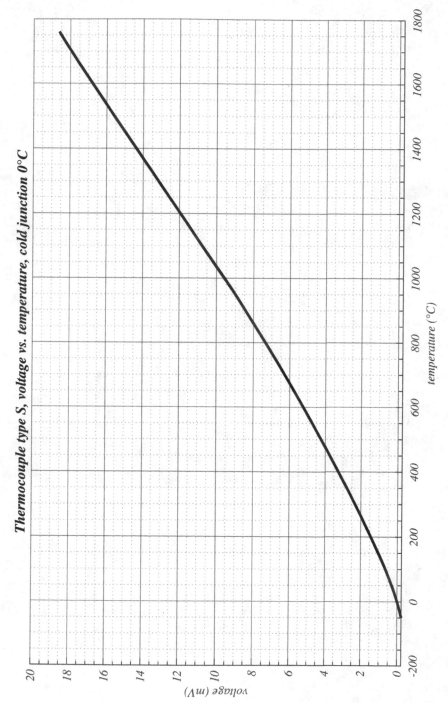

FIGURE L.1: Thermocouple type S, voltage vs. temperature, cold junction 0°C, based on data from [83]

M

Material emissivity tables

TABLE M.1: Material emissivity table

Surface material	Emissivity	at temperature	reference
Adobe	0.90	20	[84]
Asphalt, Pavement	0.93	38	[84]
Asphalt, Tar paper	0.93	20	[84]
Basalt	0.72	20	[84]
Board	0.96	38	[84]
Cadmium	0.02	25	[84]
Carborundum	0.92	1010	[84]
Chromium	0.08	38	[84]
Chromium	0.26	538	[84]
Chromium Polished	0.06	150	[84]
Clay	0.39	20	[84]
Clay Fired	0.91	70	[84]
Clay Shale	0.69	20	[84]
Clay Tiles, Dark Purple	0.78	1371–2760	[84]
Clay Tiles, Light Red	0.32–0.34	1371–2760	[84]
Clay Tiles, Red	0.40–0.51	1371–2760	[84]
Cobalt, Unoxidized	0.13	500	[84]
Cobalt, Unoxidized	0.23	1000	[84]
Columbium, Unoxidized	0.19	816	[84]
Columbium, Unoxidized	0.24	1093	[84]
Cotton Cloth	0.77	20	[84]
Cu-Zn, Brass Oxidized	0.61	200	[84]
Cu-Zn, Brass Oxidized	0.60	400	[84]
Cu-Zn, Brass Oxidized	0.61	600	[84]
Dolomite Lime	0.41	20	[84]
Emery Corundum	0.86	80	[84]
Fire Brick	0.75–0.80	1000	[84]
Granite	0.45	21	[84]
Gravel	0.28	38	[84]
Gray Brick	0.75	1100	[84]
Gum Varnish (2 coats)	0.53	21	[84]
Gum Varnish (3 coats)	0.50	21	[84]
Gypsum	0.80–0.90	20	[84]
Ice Rough	0.96	0	[84]
Ice, Smooth	0.97	0	[84]
Inconel Sheet	0.42	649	[84]
Inconel Sheet	0.58	760	[84]
Light Buff	0.80	538	[84]
Lime Clay	0.43	1371	[84]
Lime Mortar	0.90–0.92	38–260	[84]

Continues on next page

Table M.1 continued from previous page

Surface material	Emissivity	at temperature	reference
Magnesite, Refractory	0.38	1000	[84]
Magnesium	0.07–0.13	38–260	[84]
Magnesium Oxide	0.16–0.20	1027–1727	[84]
Marble, Polished Gray	0.75	38	[84]
Marble, Smooth, White	0.56	38	[84]
Marble, White	0.95	38	[84]
Mercury	0.09	0	[84]
Mercury	0.10	25	[84]
Mercury	0.10	38	[84]
Mercury	0.12	100	[84]
Molybdenum	0.06	38	[84]
Molybdenum	0.08	260	[84]
Molybdenum	0.11	538	[84]
Molybdenum	0.18	1093	[84]
Molybdenum Oxidized at 538°C	0.80	316	[84]
Molybdenum Oxidized at 538°C	0.84	371	[84]
Molybdenum Oxidized at 538°C	0.84	427	[84]
Molybdenum Oxidized at 538°C	0.83	482	[84]
Molybdenum Oxidized at 538°C	0.82	538	[84]
Monel, Ni-Cu	0.41	200	[84]
Monel, Ni-Cu	0.44	400	[84]
Monel, Ni-Cu	0.46	600	[84]
Monel, Ni-Cu Oxidized	0.43	20	[84]
Monel, Ni-Cu Oxidized at 599°C	0.46	599	[84]
Nickel Oxide	0.59–0.86	538–1093	[84]
Paints, Aluminum	0.27–0.67	38	[84]
Platinum	0.05	38	[84]
Platinum	0.05	260	[84]
Platinum	0.10	538	[84]
Platinum Black	0.93	38	[84]
Platinum Black	0.96	260	[84]
Platinum Black	0.97	1093	[84]
Platinum Black Oxidized at 593°C	0.07	260	[84]
Platinum Black Oxidized at 593°C	0.11	538	[84]
Quartz, Rough, Fused	0.93	21	[84]
Red Lead	0.93	100	[84]
Rubber, Hard	0.94	23	[84]
Rubber, Soft, Gray	0.86	24	[84]
Sand	0.76	20	[84]
Sandlime	0.59–0.63	1371–2760	[84]
Sandstone	0.67	38	[84]
Sandstone Red	0.60–0.83	38	[84]
Sawdust	0.75	20	[84]
Shale	0.69	20	[84]
Silica Glazed	0.85	1000	[84]
Silica Unglazed	0.75	1100	[84]
Silica, Glazed	0.88	1093	[84]
Silica, Unglazed	0.80	1093	[84]
Silicon Carbide	0.83–0.96	149–649	[84]
Silk Cloth	0.78	20	[84]
Slate	0.67–0.80	38	[84]
Snow Granular	0.89	-8	[84]
Snow, Fine Particles	0.82	-7	[84]

Continues on next page

Table M.1 continued from previous page

Surface material	Emissivity	at temperature	reference
Stellite, Polished	0.18	20	[84]
Stonework	0.93	38	[84]
Tin, Unoxidized	0.04	25	[84]
Tin, Unoxidized	0.05	100	[84]
Tinned Iron Bright	0.08	100	[84]
Tinned Iron, Bright	0.05	24	[84]
Brass, Unoxidized	0.04	25	[84]
Brass, Unoxidized	0.04	100	[84]
Uranium Oxide	0.79	1027	[84]
Water	0.67	38	[84]
Waterglass	0.96	20	[84]
20Ni-25Cr-55Fe, Oxidized	0.90	200	[85]
20Ni-25Cr-55Fe, Oxidized	0.97	500	[85]
3M Nextel101-C10	0.98	0–300	[85]
60Ni-12Cr-28Fe, Oxidized	0.89	270	[85]
60Ni-12Cr-28Fe, Oxidized	0.82	560	[85]
80Ni-20Cr, Oxidized	0.87	100	[85]
80Ni-20Cr, Oxidized	0.87	600	[85]
80Ni-20Cr, Oxidized	0.89	1300	[85]
A1 Lacquer, Varnish binder on rough plate	0.39	21	[85]
A1 Paint after heating to 326°C	0.35	150–316	[85]
Alloy 24ST Polished	0.09	300 K	[86]
Alloys 20-Ni, 24-CR, 55-FE, Oxidized	0.90	200	[84]
Alloys 20-Ni, 24-CR, 55-FE, Oxidized	0.97	500	[84]
Alloys 60-Ni, 12-CR, 28-FE, Oxidized	0.89	270	[84]
Alloys 60-Ni, 12-CR, 28-FE, Oxidized	0.82	560	[84]
Alloys 80-Ni, 20-CR, Oxidized	0.87	100	[84]
Alloys 80-Ni, 20-CR, Oxidized	0.87	600	[84]
Alloys 80-Ni, 20-CR, Oxidized	0.89	1300	[84]
Alum. Paint	0.55	0–100	[85]
Alumina, Flame sprayed	0.8	300 K	[86]
Aluminium Alloy 1100-0	0.05	93–427	[84]
Aluminium Alloy 24ST	0.09	24	[84]
Aluminium Alloy 24ST Polished	0.09	24	[84]
Aluminium Alloy 75ST	0.11	24	[84]
Aluminium Alloy 75ST Polished	0.08	24	[84]
Aluminium Alloy A3003, Oxidized	0.40	316	[84]
Aluminium Alloy A3003, Oxidized	0.40	482	[84]
Aluminium Bright Rolled Plate	0.04	170	[84]
Aluminium Bright Rolled Plate	0.05	500	[84]
Aluminium Commercial Sheet	0.09	100	[84]
Aluminium Heavily Oxidized	0.20	93	[84]
Aluminium Heavily Oxidized	0.31	504	[84]
Aluminium Highly Polished	0.09	100	[84]
Aluminium Highly Polished Plate	0.04	227	[84]
Aluminium Highly Polished Plate	0.06	577	[84]
Aluminium Oxidized	0.11	199	[84]
Aluminium Oxidized	0.19	599	[84]
Aluminium Oxidized at 599°C	0.11	199	[84]
Aluminium Oxidized at 599°C	0.19	599	[84]
Aluminium Roughly Polished	0.18	100	[84]
Aluminium Unoxidized	0.02	25	[84]
Aluminium Unoxidized	0.03	100	[84]

Continues on next page

Table M.1 continued from previous page

Surface material	Emissivity	at temperature	reference
Aluminium Unoxidized	0.06	500	[84]
Aluminum Anodized	0.77	300 K	[86]
Aluminum Anodized Sheet, Chromic Acid Proc	0.55	100	[85]
Aluminum Commercial sheet	0.09	300 K	[86]
Aluminum Commercial Sheet	0.09	300 K	[86]
Aluminum Commercial Sheet	0.090	100	[85]
Aluminum Foil	0.04	300 K	[86]
Aluminum Heavily Oxidized	0.2–0.31	300 K	[86]
Aluminum Heavily Oxidized	0.2–0.31	93–504	[85]
Aluminum Highly Polished	0.039–0.057	300 K	[86]
Aluminum Highly Polished	0.04–0.06	50–500	[85]
Aluminum Oxide	0.42–0.26	500–827	[85]
Aluminum Oxidized	0.11	200	[85]
Aluminum Oxidized	0.19	600	[85]
Aluminum Paint with silicone vehicle paint on Inconel	0.290	260	[85]
Aluminum Paint	0.27–0.67	300 K	[86]
Aluminum Paints & Lacquers 10% Al 22% lacquer body, on rough or smooth surface	0.520	100	[85]
Aluminum Polished	0.095	100	[85]
Aluminum Rough	0.07	300 K	[86]
Aluminum Unoxidized	0.022	25	[85]
Aluminum Unoxidized	0.028	100	[85]
Aluminum Unoxidized	0.060	500	[85]
Aluminum Polished	0.05	0°C	[87]
Aluminum Rough	0.07	0°C	[87]
Aluminum, Strongly oxidized	0.25	0°C	[87]
Antimony, Polished	0.28–0.31	300 K	[86]
Asbestos Board	0.96	0°C	[87]
Asbestos Board	0.96	20	[85]
Asbestos Board	0.96	300 K	[86]
Asbestos Cement	0.96	0–200	[85]
Asbestos Cement	0.96	0–200	[84]
Asbestos Cement Red	0.67	1371	[84]
Asbestos Cement White	0.65	1371	[84]
Asbestos Cloth	0.90	93	[85]
Asbestos Cloth	0.90	93	[84]
Asbestos Fabric	0.78	0°C	[87]
Asbestos Paper	0.93–0.945	300 K	[86]
Asbestos Paper	0.94	0°C	[87]
Asbestos Paper	0.95	0–100	[85]
Asbestos Paper	0.93	38–371	[84]
Asbestos Slate	0.96	0°C	[87]
Asbestos Slate	0.97	20	[84]
Asphalt	0.93	300 K	[86]
Asphalt	0.90–0.98	Ambient	[85]
Asphalt oil, on polished metal .001" Thick	0.27	Ambient	[85]
Asphalt oil, on polished metal .002" Thick	0.46	Ambient	[85]
Asphalt oil, on polished metal .005" Thick	0.72	Ambient	[85]
Basalt	0.72	300 K	[86]
Beryllium	0.18	300 K	[86]
Beryllium, Anodized	0.9	300 K	[86]

Continues on next page

Table M.1 continued from previous page

Surface material	Emissivity	at temperature	reference
Bismuth, Bright	0.34	300 K	[86]
Bismuth, Bright Bismuth, Unoxidized	0.05	25	[84]
Bismuth, Bright Bismuth, Unoxidized	0.06	100	[84]
Bismuth, Unoxidized	0.048	25	[85]
Bismuth, Unoxidized	0.061	100	[85]
Black Enamel Paint	0.8	300 K	[86]
Black Epoxy Paint	0.89	300 K	[86]
Black Glass Paint	0.90	0–100	[85]
Black lacquer on iron	0.875	300 K	[86]
Black Matte shellac	0.910	77–146	[85]
Black on White Lacquer	0.800–0.950	38–93	[85]
Black Parson Optical	0.95	300 K	[86]
Black Shiny Lacquer, Sprayed on iron	0.875	24	[85]
Black Shiny shellac on tinned iron sheet	0.821	21	[85]
Black Silicone Paint	0.93	300 K	[86]
Brass 62%Cu.37%Zn. Polished	0.03	257	[84]
Brass 62%Cu.37%Zn. Polished	0.04	377	[84]
Brass 73%Cu.27%Zn. Polished	0.03	247	[84]
Brass 73%Cu.27%Zn. Polished	0.03	357	[84]
Brass 83%Cu.17%Zn. Polished	0.03	277	[84]
Brass Burnished to Brown	0.40	20	[84]
Brass Dull Plate	0.22	300 K	[86]
Brass Matte	0.07	20	[84]
Brass Oxidized	0.61	200	[85]
Brass Oxidized	0.59	600	[85]
Brass Oxidized 600oC	0.6	300 K	[86]
Brass Polished	0.03	300 K	[86]
Brass Polished	0.03	200	[85]
Brass Rolled Plate Natural Surface	0.06	300 K	[86]
Brass Rolled Sheet	0.06	20	[85]
Brass Unoxidized	0.035	25	[85]
Brass Unoxidized	0.035	100	[85]
Brass, Dull, Tarnished	0.22	0°C	[87]
Brass, Polished	0.03	0°C	[87]
Brick Building	0.450	1000	[85]
Brick Fire	0.750	1000	[85]
Brick Fire Clay	0.75	1371	[84]
Brick Gault Cream	0.26–0.30	1371–2760	[84]
Brick Grog, Brick, Glazed	0.750	1100	[85]
Brick Red, Rough	0.93	21	[84]
Brick Red, Rough, No gross irregularities	0.930	20	[85]
Brick Silica	0.80	1000	[85]
Brick Silica	0.85	1100	[85]
Brick, Common	0.85	0°C	[87]
Brick, Fireclay	0.75	300 K	[86]
Brick, Glazed, Rough	0.85	0°C	[87]
Brick, Red rough	0.93	300 K	[86]
Brick, Refractory, Rough	0.94	0°C	[87]
Bronze Paint	0.80	0–100	[85]
Bronze, Polished	0.10	0°C	[87]
Bronze, Polished	0.10	50	[85]
Bronze, Porous, Rough	0.55	0°C	[87]
Cadmium	0.02	300 K	[86]

Continues on next page

Table M.1 continued from previous page

Surface material	Emissivity	at temperature	reference
Carbon Candle Soot	0.952	97–271	[85]
Carbon Candle Soot	0.95	121	[84]
Carbon Filament	0.77	300 K	[86]
Carbon Filament	0.53	1000–1400	[85]
Carbon Filament	0.95	260	[84]
Carbon Graphite	0.70–0.80	0–3600	[85]
Carbon Graphite, Pressed, Filed surface	0.980	250–510	[85]
Carbon Graphitized	0.76	100	[84]
Carbon Graphitized	0.75	300	[84]
Carbon Graphitized	0.71	500	[84]
Carbon Lamp, Black, Water glass coating	0.96	20–400	[85]
Carbon Lampblack	0.95	25	[84]
Carbon Pressed Filled Surface	0.98	300 K	[86]
Carbon Soot applied to solid	0.96	50–1000	[85]
Carbon Soot with water glass	0.96	20–200	[85]
Carbon Unoxidized	0.81	25	[85]
Carbon Unoxidized	0.81	100	[85]
Carbon Unoxidized	0.81	500	[85]
Carbon Unoxidized	0.81	25	[84]
Carbon Unoxidized	0.81	100	[84]
Carbon Unoxidized	0.79	500	[84]
Carbon, Not oxidized	0.81	300 K	[86]
Carbon, Purified	0.80	0°C	[87]
Carborundum 87SiC; 2.3 density	0.92–0.82	1010–1400	[85]
Cast Iron Liquid	0.29	1535	[84]
Cast Iron Oxidized	0.64	199	[84]
Cast Iron Oxidized	0.78	599	[84]
Cast Iron Stong Oxidation	0.95	40	[84]
Cast Iron Strong Oxidation	0.95	250	[84]
Cast Iron Unoxidized	0.21	100	[84]
Cast Iron, Newly turned	0.44	300 K	[86]
Cast iron, Polished	0.21	0°C	[87]
Cast iron, Rough casting	0.81	0°C	[87]
Cast Iron, Turned and heated	0.60–0.70	300 K	[86]
Cement	0.54	300 K	[86]
Ceramic Alumina on Inconel	0.69–0.45	427–1093	[84]
Ceramic Coating No. C20A	0.73–0.87	93–399	[84]
Ceramic Earthenware	0.90	20	[85]
Ceramic Earthenware, Glazed	0.90	21	[84]
Ceramic Earthenware, Matte	0.93	21	[84]
Ceramic Greens No. 5210–2C	0.89–0.82	93–399	[84]
Ceramic Porcelain	0.92	22	[84]
Ceramic Porcelain, Glazed	0.92	20	[85]
Ceramic Refractory Black	0.94	93	[85]
Ceramic Refractory White	0.90	93	[85]
Ceramic White Aluminium Oxide	0.90	93	[84]
Ceramic Zirconia on Inconel	0.62–0.45	427–1093	[84]
Charcoal, Powdered	0.96	0°C	[87]
Chromium Oxidized	0.08	316	[85]
Chromium Oxidized	0.18	482	[85]
Chromium Oxidized	0.27	650	[85]
Chromium Oxidized	0.36	816	[85]
Chromium Oxidized	0.66	982	[85]

Continues on next page

Table M.1 continued from previous page

Surface material	Emissivity	at temperature	reference
Chromium polished	0.058	300 K	[86]
Chromium Polished	0.10	50	[85]
Chromium Polished	0.28–0.38	500–1000	[85]
Chromium Unoxidized	0.08	100	[85]
Chromium, Polished	0.10	0°C	[87]
Chromnickel	0.640–0.760	52–1034	[85]
Clay	0.91	300 K	[86]
Clay, Fired	0.91	0°C	[87]
Clear Silicone Vehicle Coating 0.001–0.150" thick On Al Alloys, 24ST, 75ST	0.770, 0.820	260	[85]
Clear Silicone Vehicle Coating 0.001–0.150" thick On Dow Metal	0.740	260	[85]
Clear Silicone Vehicle Coating 0.001–0.150" thick On mild steels	0.660	260	[85]
Clear Silicone Vehicle Coating 0.001–0.150" thick On stainless steels 316, 301, 347	0.680, 0.750, 0.750	260	[85]
Coal	0.8	300 K	[86]
Cobalt, Unoxidized	0.13	500	[85]
Cobalt, Unoxidized	0.23	1000	[85]
Columbium Oxidized	0.73	816	[85]
Columbium Oxidized	0.70	927	[85]
Columbium Polished	0.19	1500	[85]
Columbium Polished	0.24	2000	[85]
Concrete	0.85	300 K	[86]
Concrete	0.54	0°C	[87]
Concrete	0.94	0–100	[85]
Concrete Rough	0.94	0–1093	[84]
Concrete Tiles	0.63	300 K	[86]
Concrete Tiles	0.630	1000	[85]
Concrete Tiles Black	0.94–0.91	1371–2760	[84]
Concrete Tiles, Brown	0.87–0.83	1371–2760	[84]
Concrete Tiles, Natural	0.63–0.62	1371–2760	[84]
Concrete, Rough	0.94	300 K	[86]
Copper Black, Oxidized	0.78	38	[84]
Copper Calorized	0.26	100	[85]
Copper Calorized, Oxidized	0.18	200	[85]
Copper Calorized, Oxidized	0.19	600	[85]
Copper Commercial, Scoured to a shine	0.07	20	[85]
Copper Cuprous Oxide	0.66–0.54	800–1100	[85]
Copper Cuprous Oxide	0.87	38	[84]
Copper Cuprous Oxide	0.83	260	[84]
Copper Cuprous Oxide	0.77	538	[84]
Copper Dow Metal	0.15	(18)-316	[84]
Copper Electroplated	0.03	300 K	[86]
Copper Etched	0.09	38	[84]
Copper Heated and covered with thick oxide layer	0.78	300 K	[86]
Copper Highly Polished	0.02	38	[84]
Copper Liquid	0.15		[85]
Copper Matte	0.22	38	[84]
Copper Molten	0.15	538	[84]

Continues on next page

Table M.1 continued from previous page

Surface material	Emissivity	at temperature	reference
Copper Molten	0.16	1077	[84]
Copper Molten	0.13	1221	[84]
Copper Nickel Alloy, Polished	0.059	300 K	[86]
Copper Nickel Plated	0.37	38–260	[84]
Copper Oxidized	0.6–0.7	50	[85]
Copper Oxidized	0.60	200	[85]
Copper Oxidized	0.88	500	[85]
Copper Plate, Heated at 600°C	0.570	200–600	[85]
Copper Polished	0.023–0.052	300 K	[86]
Copper Polished	0.02–0.05	50–100	[85]
Copper Polished	0.03	38	[84]
Copper Rolled	0.64	38	[84]
Copper Rough	0.74	38	[84]
Copper Roughly Polished	0.07	38	[84]
Copper Unoxidized	0.02	100	[85]
Copper with thick oxide layer	0.78	25	[85]
Copper, Commercial burnished	0.07	0°C	[87]
Copper, Oxidized	0.65	0°C	[87]
Copper, Oxidized to black	0.88	0°C	[87]
Copper, Polished	0.01	0°C	[87]
Cotton Cloth	0.77	300 K	[86]
Covex D Glass	0.76	320	[85]
Dow Metal	0.24–0.20	232–400	[85]
Dull Black Varnish	0.80–0.95	40–100	[85]
Electrical tape, Black plastic	0.95	0°C	[87]
Enamel **	0.90	0°C	[87]
Enamel, White, Fused on Iron	0.900	19	[85]
Flat Black Lacquer	0.960–0.980	38–93	[85]
Formica	0.93	0°C	[87]
Frozen Soil	0.93	0°C	[87]
Fused Quartz	0.75	320	[85]
Glass	0.92	0°C	[87]
Glass Convex D	0.80	100	[84]
Glass Convex D	0.80	316	[84]
Glass Convex D	0.76	500	[84]
Glass Nonex	0.82	100	[84]
Glass Nonex	0.82	316	[84]
Glass Nonex	0.78	500	[84]
Glass smooth	0.92–0.94	300 K	[86]
Glass Smooth	0.95	0–200	[85]
Glass Smooth	0.87–0.72	250–1000	[85]
Glass Smooth	0.70–0.67	1100–1500	[85]
Glass Smooth	0.92–0.94	0–93	[84]
Glass, Frosted	0.96	0°C	[87]
Glass, Pyrex	0.85–0.95	300 K	[86]
Glossy Black Varnish sprayed on iron	0.87	20	[85]
Gold Carefully Polished	0.02–0.03	200–600	[85]
Gold Enamel	0.37	100	[85]
Gold Enamel	0.37	0–100	[85]
Gold Enamel	0.37	100	[84]
Gold not polished	0.47	300 K	[86]
Gold on .0005 Nickel	0.07–0.09	93–399	[84]
Gold on .0005 Silver	0.11–0.14	93–399	[84]

Continues on next page

Table M.1 continued from previous page

Surface material	Emissivity	at temperature	reference
Gold Polished	0.025	300 K	[86]
Gold Polished	0.02	38–260	[84]
Gold Polished	0.03	538–1093	[84]
Gold Pure, Highly polished	0.02	100	[85]
Gold Unoxidized	0.02	100	[85]
Gold Unoxidized	0.03	500	[85]
Gold Polished	0.02	0°C	[87]
Granite	0.45	300 K	[86]
Graphite	0.70–0.80	0–3600	[85]
Gravel	0.28	300 K	[86]
Gray Paint	0.95	0–100	[85]
Green Paint	0.95	0–100	[85]
Gypsum	0.85	300 K	[86]
Gypsum 0.02" thick on smooth or blackened plate	0.93	20	[85]
Haynes Alloy C, Oxidized Haynes Alloy 25, Oxidized	0.86–0.89	316–1093	[84]
Haynes Alloy C, Oxidized Haynes Alloy X, Oxidized	0.85–0.88	316–1093	[84]
Human Skin	0.985	36–7-37.2	[85]
Ice	0.97	0°C	[87]
Ice rough	0.985	300 K	[86]
Ice smooth	0.966	300 K	[86]
Inconel Sheet Inconel B, Polished	0.21	24	[84]
Inconel Sheet Inconel X, Polished	0.19	24	[84]
Inconel Type B	0.350–0.550	450–1620	[85]
Inconel Type X	0.550–0.780		[85]
Inconel X Oxidized	0.71	300 K	[86]
Iron Cast Plate, Oxidized, Rough	0.82	23	[85]
Iron Cast Plate, Oxidized, Smooth	0.8	23	[85]
Iron Liquid	0.42–0.45	1516–1771	[84]
Iron Liquid Unoxidized	0.29	–	[85]
Iron Molten Armco	0.400–0.410	1521–1689	[85]
Iron Molten Pure	0.420–0.450	1516–1771	[85]
Iron Newly Turned	0.440	22	[85]
Iron Oxide	0.85–0.89	500–1200	[85]
Iron Oxidized	0.64–0.78	200–600	[85]
Iron Oxidized	0.74	100	[85]
Iron Oxidized	0.84	500	[85]
Iron Oxidized	0.74	100	[84]
Iron Oxidized	0.84	499	[84]
Iron Oxidized	0.89	1199	[84]
Iron Plate, Completely rusted	0.690	19	[85]
Iron Plate, Pickled, Then rusted red	0.610	20	[85]
Iron Polished	0.14–0.38	300 K	[86]
Iron Polished	0.210	200	[85]
Iron Red Rust	0.70	25	[84]
Iron Rough-ingot	0.870–0.950	927–1116	[85]
Iron Rusted	0.65	25	[85]
Iron Rusted	0.65	25	[84]
Iron Smooth Oxidized electrolytic iron	0.780–0.820	127–527	[85]
Iron Strongly Oxidized	0.95	40	[85]
Iron Strongly Oxidized	0.95	250	[85]

Continues on next page

Surface material	Emissivity	at temperature	reference
Iron Turned and Heated	0.600–0.700	882–990	[85]
Iron Unoxidized	0.21	100	[85]
Iron Unoxidized	0.89	1200	[85]
Iron Unoxidized	0.05	100	[84]
Iron Wrought, Dull	0.50	100	[85]
Iron Wrought, Dull oxidized	0.940	21–360	[85]
Iron Wrought, Highly polished	0.280	38–250	[85]
Iron, Dark gray surface	0.31	300 K	[86]
Iron, Hot rolled	0.77	0°C	[87]
Iron, Oxidized	0.74	0°C	[87]
Iron, Plate rusted red	0.61	300 K	[86]
Iron, Rough ingot	0.87–0.95	300 K	[86]
Iron, Sheet galvanized, Burnished	0.23	0°C	[87]
Iron, Sheet, Galvanized, Oxidized	0.28	0°C	[87]
Iron, Shiny, Etched	0.16	0°C	[87]
Iron, Wrought, Polished	0.28	0°C	[87]
Lacquer Black	0.96	93	[84]
Lacquer Blue, On Aluminum Foil	0.78	38	[84]
Lacquer Clear, On Aluminum Foil (2 coat)	0.08(0.09)	93	[84]
Lacquer Clear, On Bright Copper	0.66	93	[84]
Lacquer Clear, On Tarnished Copper	0.64	93	[84]
Lacquer coatings, 0.001–0.015" thick on Alum. Alloys	0.870–0.970	38–150	[85]
Lacquer Red, On Aluminum Foil (2 coat)	0.61(0.74)	38	[84]
Lacquer White	0.95	93	[84]
Lacquer White, On Aluminum Foil (2 coat)	0.69(0.88)	38	[84]
Lacquer Yellow, On Aluminum Foil (2 coat)	0.57(0.79)	38	[84]
Lacquer, Bakelite	0.93	0°C	[87]
Lacquer, Black, Dull	0.97	0°C	[87]
Lacquer, Black, Shiny	0.87	0°C	[87]
Lacquer, White	0.87	0°C	[87]
Lamp Black	0.95	0–100	[85]
Lampblack	0.96	0°C	[87]
Lampblack paint	0.96	300 K	[86]
Lead Gray Oxidized	0.28	38	[84]
Lead Oxidized	0.43	300 K	[86]
Lead Oxidized	0.63	200	[85]
Lead Oxidized	0.43	38	[84]
Lead Oxidized at 593°C	0.63	38	[84]
Lead Oxidized, Gray	0.280	24	[85]
Lead Polished	0.06–0.08	38–260	[84]
Lead Pure (99.96%) Unoxidized	0.057–0.075	127–227	[85]
Lead pure Unoxidized	0.057–0.075	300 K	[86]
Lead Rough	0.43	38	[84]
Lead, Gray	0.28	0°C	[87]
Lead, Oxidized	0.63	0°C	[87]
Lead, Red, Powdered	0.93	0°C	[87]
Lead, Shiny	0.08	0°C	[87]
Lime Wash	0.91	300 K	[86]
Limestone	0.90–0.93	300 K	[86]
Magnesia	0.72	300 K	[86]

Continues on next page

Table M.1 continued from previous page

Surface material	Emissivity	at temperature	reference
Magnesite	0.38	300 K	[86]
Magnesite Refractory Brick	0.380	1000	[85]
Magnesium Oxide	0.20–0.55	300 K	[86]
Magnesium Oxide	0.550–0.200	227–826	[85]
Magnesium Oxide	0.200	900–1704	[85]
Magnesium Polished	0.07–0.13	300 K	[86]
Marble White	0.95	300 K	[86]
Marble, Light Grey Polished	0.903	0–100	[85]
Masonry Plastered	0.93	300 K	[86]
Mercury liquid	0.1	300 K	[86]
Mercury, Pure	0.10	0°C	[87]
Mercury, Unoxidized	0.09	0	[85]
Mercury, Unoxidized	0.10	25	[85]
Mercury, Unoxidized	0.12	100	[85]
Mikron High Temp Test Paint (Spirex SP102)	0.999	Ambient-650	[85]
Mild Steel	0.20–0.32	300 K	[86]
Molybdenum Filament	0.096–0.202	827–2593	[85]
Molybdenum Oxidized	0.82	538	[85]
Molybdenum polished	0.05–0.18	300 K	[86]
Molybdenum Polished	0.05	538	[85]
Molybdenum Polished	0.17	1482	[85]
Molybdenum Unoxidized	0.13	1000	[85]
Molybdenum Unoxidized	0.19	1500	[85]
Molybdenum Unoxidized	0.24	2000	[85]
Monel Metal, Oxidized	0.43	200	[85]
Monel Metal, Oxidized	0.43	600	[85]
Monel, Ni-Cu Type 17-7PH Polished	0.09–0.16	149–816	[84]
Monel, Ni-Cu Type C1020, Oxidized	0.87–0.91	316–1093	[84]
Monel, Ni-Cu Type PH-15-7 MO	0.07–0.19	149–649	[84]
Mortar	0.87	300 K	[86]
Nichrome Oxidized	0.95–0.98	50–500	[85]
Nichrome Wire Clean	0.65	50	[85]
Nichrome Wire Clean	0.71–0.79	500–1000	[85]
Nichrome Wire, Bright	0.65–0.79	300 K	[86]
Nickel Electrolytic	0.04	38	[84]
Nickel Electrolytic	0.06	260	[84]
Nickel Electrolytic	0.10	538	[84]
Nickel Electrolytic	0.16	1093	[84]
Nickel Electroplated, Not Polished	0.110	20	[85]
Nickel Electroplated, Polished	0.045	23	[85]
Nickel Oxide	0.590–0.860	650–1254	[85]
Nickel Oxidized	0.37	200	[85]
Nickel Oxidized	0.85	871	[85]
Nickel Oxidized	0.85	1200	[85]
Nickel Oxidized	0.31–0.46	38–260	[84]
Nickel Plate, Oxidized by heating at 600°C	0.370–0.480	200–600	[85]
Nickel Polished	0.12	low	[85]
Nickel Polished	0.32	1204	[85]
Nickel Polished	0.05	38	[84]
Nickel Unoxidized	0.045	25	[85]
Nickel Unoxidized	0.06	100	[85]
Nickel Unoxidized	0.12	500	[85]

Continues on next page

Table M.1 continued from previous page

Surface material	Emissivity	at temperature	reference
Nickel Unoxidized	0.19	1000	[85]
Nickel Unoxidized	0.05	25	[84]
Nickel Unoxidized	0.06	100	[84]
Nickel Unoxidized	0.12	500	[84]
Nickel Unoxidized	0.19	1000	[84]
Nickel Wire	0.096–0.186	187–1007	[85]
Nickel, Elctroplated	0.03	300 K	[86]
Nickel, On cast iron	0.05	0°C	[87]
Nickel, Oxidized	0.59–0.86	300 K	[86]
Nickel, Polished	0.072	300 K	[86]
Nickel, Pure polished	0.05	0°C	[87]
Nickel-Silver Polished	0.135	100	[85]
Nonex Glass	0.82	320	[85]
Oak, Planed	0.89	300 K	[86]
Oak, Planed	0.900	21	[85]
Oil Layers (Linseed Oil) on Aluminum Foil	0.087	100	[85]
Oil Layers (Linseed Oil) on Aluminum Foil +1, 2 coats oil	0.561–0.574	100	[85]
Oil Paints, 16 diff. (all colors)	0.920–0.960	100	[85]
Oil paints, All colors	0.92–0.96	300 K	[86]
Oil, Linseed On Aluminum Foil, 1 coat	0.59	121	[84]
Oil, Linseed On Aluminum Foil, 2 coats	0.51	121	[84]
Oil, Linseed On Aluminum Foil, Uncoated	0.09	121	[84]
Oil, Linseed On Polished Iron, .001 Film	0.22	38	[84]
Oil, Linseed On Polished Iron, .002 Film	0.45	38	[84]
Oil, Linseed On Polished Iron, .004 Film	0.65	38	[84]
Oil, Linseed On Polished Iron, Thick Film	0.83	38	[84]
Other A1 paints, Varying age and Al content	0.270–0.670	100	[85]
Paint	0.96	300 K	[86]
Paint, Oil, Average	0.94	0°C	[87]
Paint, Silver finish**	0.31	0°C	[87]
Paints Black, CuO	0.96	24	[84]
Paints Blue, Cu2-O3	0.94	24	[84]
Paints Green, Cu2O3	0.92	24	[84]
Paints Red, Fe2O3	0.91	24	[84]
Paints White Al2O3	0.94	24	[84]
Paints White MgCO3	0.91	24	[84]
Paints White MgO2	0.91	24	[84]
Paints White PbCO3	0.93	24	[84]
Paints White ThO2	0.90	24	[84]
Paints White Y2O3	0.90	24	[84]
Paints White ZnO	0.95	24	[84]
Paints White, ZrO2	0.95	24	[84]
Paints Yellow PbCrO4	0.93	24	[84]
Paints Yellow, PbO	0.90	24	[84]
Paints, Aluminum 10% Al	0.52	38	[84]
Paints, Aluminum 20% Al	0.30	38	[84]
Paints, Aluminum Dow XP-310	0.22	93	[84]
Paints, Oil All colors	0.92–0.96	93	[84]
Paints, Oil Black	0.92	93	[84]
Paints, Oil Black Gloss	0.30	21	[84]
Paints, Oil Camouflage Green	0.85	52	[84]

Continues on next page

Table M.1 continued from previous page

Surface material	Emissivity	at temperature	reference
Paints, Oil Flat Black	0.88	27	[84]
Paints, Oil Flat White	0.91	27	[84]
Paints, Oil Gray-Green	0.95	21	[84]
Paints, Oil Green	0.95	93	[84]
Paints, Oil Lamp Black	0.96	98	[84]
Paints, Oil Red	0.95	93	[84]
Paints, Oil White	0.94	93	[84]
Palladium Plate (.00005 on .0005 silver)	0.16–0.17	93–399	[84]
Paper	0.93	300 K	[86]
Paper offset	0.55	300 K	[86]
Paper Thinipasted on Tinned or Blackened Plate	0.92–0.94	19	[85]
Paper, Any Color	0.94	0–100	[85]
Paper, Black, Dull	0.94	0°C	[87]
Paper, Black, Shiny	0.90	0°C	[87]
Paper, White	0.90	0°C	[87]
Pine	0.84	300 K	[86]
Plaster	0.98	300 K	[86]
Plaster	0.91	0–200	[85]
Plaster Board	0.91	300 K	[86]
Plaster, Rough	0.91	300 K	[86]
Plastics	0.90–0.97	300 K	[86]
Plastics, Opaque any color	0.95	25	[85]
Platinum Cleaned Polished	0.05–0.10	200–600	[85]
Platinum Filament	0.036–0.192	27–1227	[85]
Platinum Unoxidized	0.037	25	[85]
Platinum Unoxidized	0.047	100	[85]
Platinum Unoxidized	0.096	500	[85]
Platinum Unoxidized	0.152	1000	[85]
Platinum Unoxidized	0.191	1500	[85]
Platinum Wire	0.06–0.07	50–200	[85]
Platinum Wire	0.10–0.16	500–1000	[85]
Platinum, Polished plate	0.054–0.104	300 K	[86]
Platinum, Pure, Polished	0.08	0°C	[87]
Polypropylene	0.97	300 K	[86]
Polytetrafluoroethylene (PTFE)	0.92	300 K	[86]
Porcelain glazed	0.93	300 K	[86]
Porcelain, Glazed	0.92	300 K	[86]
Porcelain, Glazed	0.92	0°C	[87]
Propellant Liquid rocket engine	0.90	600–4500	[85]
PVC	0.91–0.93	300 K	[86]
Pyrex	0.92	300 K	[86]
Pyrex	0.90	0–300	[85]
Quartz	0.93	0°C	[87]
Quartz glass	0.93	300 K	[86]
Quartz Glass, 1.98mm Thick	0.900–0.410	282–838	[85]
Quartz Glass, 6.88mm Thick	0.930–0.470	300–838	[85]
Quartz Opaque	0.920–0.680	300–838	[85]
Quartz Rough, Fused	0.930	21	[85]
Quartz, Rough, Fused Glass, 1.96 mm	0.90	282	[84]
Quartz, Rough, Fused Glass, 1.96 mm	0.41	838	[84]
Quartz, Rough, Fused Glass, 6.88 mm	0.93	282	[84]
Quartz, Rough, Fused Glass, 6.88 mm	0.74	838	[84]

Continues on next page

Table M.1 continued from previous page

Surface material	Emissivity	at temperature	reference
Quartz, Rough, Fused Opaque	0.92	299	[84]
Quartz, Rough, Fused Opaque	0.68	838	[84]
Radiator Paint White, Cream, Bleach	0.790, 0.770, 0.840	100	[85]
Radiator Paint, Bronze	0.510	100	[85]
Rhodium Flash (.0002 on .0005 Ni)	0.10–0.18	93–371	[84]
Roofing Paper	0.91	300 K	[86]
Roofing Paper	0.910	21	[85]
Rubber	0.93	0°C	[87]
Rubber, Foam	0.9	300 K	[86]
Rubber, Hard glossy plate	0.94	300 K	[86]
Rubber, Natural hard	0.91	300 K	[86]
Rubber, Natural oft	0.86	300 K	[86]
Salt	0.34	300 K	[86]
Sand	0.76	300 K	[86]
Sandstone	0.59	300 K	[86]
Sapphire	0.48	300 K	[86]
Sawdust	0.75	300 K	[86]
Shellac, Black, Dull	0.91	0°C	[87]
Shellac, Black, Shiny	0.82	0°C	[87]
Silica	0.79	300 K	[86]
Silica (98 Si O2, Fe-free) effect of grain size, Microns 10 microns	0.420–0.330	1010–1566	[85]
Silica (98 Si O2, Fe-free) effect of grain size, Microns 70–600 microns	0.620–0.460	1010–1566	[85]
Silicon Carbide	0.83–0.96	300 K	[86]
Silver Cleaned Polished	0.02–0.03	200–600	[85]
Silver Plate (.0005 on Ni)	0.06–0.07	93–371	[84]
Silver Polished	0.02–0.03	300 K	[86]
Silver Polished	0.052	100	[85]
Silver Polished	0.01	38	[84]
Silver Polished	0.02	260	[84]
Silver Polished	0.03	538	[84]
Silver Polished	0.03	1093	[84]
Silver Unoxidized	0.02	100	[85]
Silver Unoxidized	0.035	500	[85]
Snow	0.80	0°C	[87]
Snow White Enamel varnish on rough iron plate	0.906	23	[85]
Soil	0.90–0.95	300 K	[86]
Soil Black Loam	0.66	20	[84]
Soil Plowed Field	0.38	20	[84]
Soil Surface	0.38	38	[84]
Soot Acetylene	0.97	24	[84]
Soot Camphor	0.94	24	[84]
Soot Candle	0.95	121	[84]
Soot Coal	0.95	20	[84]
Stainless Steel 18-8 Oxidized	0.83	93–371	[85]
Stainless Steel 18-8 Buffed	0.160	20	[85]
Stainless Steel 18-8 Polished	0.16	93	[85]
Stainless Steel 18-8 Polished	0.19	371	[85]
Stainless Steel 303	0.74	316	[85]
Stainless Steel 303 Oxidized	0.87	1093	[85]

Continues on next page

Table M.1 continued from previous page

Surface material	Emissivity	at temperature	reference
Stainless Steel 304 (8Cri 18Ni) light silvery, Rough brown, After heating	0.440–0.360	216–490	[85]
Stainless Steel 304 (8Cri 18Ni) light silvery, Rough brown, After 42 hours of heating at 527°C	0.620–0.730	216–527	[85]
Stainless Steel 310 (25Cr, 20Ni) Brown, Splotched, Oxidized from furnace service	0.900–0.970	216–527	[85]
Stainless Steel Allegheny metal No. 4, Polished	0.130	100	[85]
Stainless Steel Allegheny metal No. 66, Polished	0.110	100	[85]
Stainless Steel, Polished	0.075	300 K	[86]
Stainless Steel, Type 301	0.54–0.63	300 K	[86]
Stainless Steel, Weathered	0.85	300 K	[86]
Steel Alloyed (8%Ni, 18%Cr)	0.35	500	[85]
Steel Alloys Type 17-7PH	0.44–0.51	93–316	[84]
Steel Alloys Type 301, Polished	0.27	24	[84]
Steel Alloys Type 301, Polished	0.57	232	[84]
Steel Alloys Type 301, Polished	0.55	949	[84]
Steel Alloys Type 303, Oxidized	0.74–0.87	316–1093	[84]
Steel Alloys Type 310, Rolled	0.56–0.81	816–1149	[84]
Steel Alloys Type 316, Polished	0.28	24	[84]
Steel Alloys Type 316, Polished	0.57	232	[84]
Steel Alloys Type 316, Polished	0.66	949	[84]
Steel Alloys Type 321	0.27–0.32	93–427	[84]
Steel Alloys Type 321 Polished	0.18–0.49	149–816	[84]
Steel Alloys Type 321 w/BK Oxide	0.66–0.76	93–427	[84]
Steel Alloys Type 347, Oxidized	0.87–0.91	316–1093	[84]
Steel Alloys Type 350	0.18–0.27	93–427	[84]
Steel Alloys Type 350, Polished	0.11–0.35	149–982	[84]
Steel Alloys Type 446, Polished	0.15–0.37	149–816	[84]
Steel Aluminized	0.79	50–500	[85]
Steel Calorized, Oxidized	0.52	200	[85]
Steel Calorized, Oxidized	0.57	600	[85]
Steel Cast, Polished	0.52–0.56	750–1050	[85]
Steel Cold Rolled	0.75–0.85	93	[84]
Steel Dull Nickel Plated	0.11	20	[85]
Steel Flat, Rough Surface	0.95–0.98	50	[85]
Steel Galvanized New	0.23	300 K	[86]
Steel Galvanized Old	0.88	300 K	[86]
Steel Ground Sheet	0.55–0.61	938–1099	[84]
Steel Mild Steel, Liquid	0.28	1599–1799	[84]
Steel Mild Steel, Polished	0.10	24	[84]
Steel Mild Steel, Polished Smooth	0.12	24	[84]
Steel Molten	0.420–0.530	1500–1650	[85]
Steel Molten Mild	0.280	1600–1800	[85]
Steel Molten, Unoxidized	0.280	Liquid	[85]
Steel Molten, Various with 0.25–1.2% (slightly oxidized surfaces.)	0.270–0.390	1560–1710	[85]
Steel Oxidized	0.79	300 K	[86]
Steel Oxidized	0.80	25	[85]
Steel Oxidized	0.79	200	[85]
Steel Oxidized	0.79	600	[85]

Continues on next page

Table M.1 continued from previous page

Surface material	Emissivity	at temperature	reference
Steel Plate, Rough	0.94	40	[85]
Steel Plate, Rough	0.97	400	[85]
Steel Plate, Rough	0.57	600	[85]
Steel Polished	0.07	300 K	[86]
Steel Polished Sheet	0.07	38	[84]
Steel Polished Sheet	0.10	260	[84]
Steel Polished Sheet	0.14	538	[84]
Steel Sheet with Shiny layer of oxide	0.82	20	[85]
Steel Sheet, Ground	0.55–0.61	938–1100	[85]
Steel Sheet, Rolled	0.66	21	[85]
Steel Sheet, Strong, Rough Oxide Layer	0.80	24	[85]
Steel Steel Oxidized	0.80	25	[84]
Steel Steel, Unoxidized	0.08	100	[84]
Steel Unoxidized	0.08	100	[85]
Steel, Galvanized	0.28	0°C	[87]
Steel, Oxidized strongly	0.88	0°C	[87]
Steel, Rolled freshly	0.24	0°C	[87]
Steel, Rough surface	0.96	0°C	[87]
Steel, Rusty red	0.69	0°C	[87]
Steel, Sheet, Nickel plated	0.11	0°C	[87]
Steel, Sheet, Rolled	0.56	0°C	[87]
Tantalum Filament	0.19–0.31	1327–3000	[85]
Tantalum Unoxidized	0.21	1500	[85]
Tantalum Unoxidized	0.26	2000	[85]
Tantalum Unoxidized	0.14	727	[84]
Tantalum Unoxidized	0.19	1093	[84]
Tantalum Unoxidized	0.26	1982	[84]
Tantalum Unoxidized	0.30	2930	[84]
Tar paper	0.92	0°C	[87]
Thoria	0.28	300 K	[86]
Thorium Oxide	0.58–0.36	277–500	[85]
Tile	0.97	300 K	[86]
Tin Unoxidized	0.04	300 K	[86]
Tin Unoxidized	0.05	25	[85]
Tin, Burnished	0.05	0°C	[87]
Tin, Commercial tin-plated sheet iron	0.07–0.08	100	[85]
Titanium Alloy C110M, Oxidized at 538°	0.51–0.61	93–427	[84]
Titanium Alloy C110M, Polished	0.08–0.19	149–649	[84]
Titanium Alloy T1-95A Oxidized at 538°	0.35–0.48	93–427	[84]
Titanium Anodized onto SS	0.96–0.82	93–316	[84]
Titanium polished	0.19	300 K	[86]
Tungsten	0.05	0°C	[87]
Tungsten aged filament	0.032–0.35	300 K	[86]
Tungsten Filament	0.39	3316	[85]
Tungsten Filament (Aged)	0.03	38	[84]
Tungsten Filament (Aged)	0.11	538	[84]
Tungsten Filament (Aged)	0.35	2760	[84]
Tungsten Filament, Aged	0.32–0.35	27–3316	[85]
Tungsten polished	0.04	300 K	[86]
Tungsten Unoxidized	0.024	25	[85]
Tungsten Unoxidized	0.032	100	[85]
Tungsten Unoxidized	0.071	500	[85]
Tungsten Unoxidized	0.15	1000	[85]

Continues on next page

Table M.1 continued from previous page

Surface material	Emissivity	at temperature	reference
Tungsten Unoxidized	0.23	1500	[85]
Tungsten Unoxidized	0.28	2000	[85]
Tungsten Unoxidized	0.02	25	[84]
Tungsten Unoxidized	0.03	100	[84]
Tungsten Unoxidized	0.07	500	[84]
Tungsten Unoxidized	0.15	1000	[84]
Tungsten Unoxidized	0.23	1500	[84]
Tungsten Unoxidized	0.28	2000	[84]
Turbojet Engine Operating	0.900	350–600	[85]
Water	0.95–0.963	300 K	[86]
Water	0.98	0°C	[87]
Water	0.96	Ambient	[85]
White Lacquer	0.95	0–100	[85]
Wood Beech, Planed	0.94	70	[84]
Wood Beech, Planned	0.935	300 K	[86]
Wood Oak, Planed	0.89	0–200	[85]
Wood Oak, Planed	0.91	38	[84]
Wood Oak, planned	0.885	300 K	[86]
Wood Spruce, Sanded	0.82	93	[85]
Wood Spruce, Sanded	0.89	38	[84]
Wood, Pine	0.95	300 K	[86]
Wrought Iron	0.94	300 K	[86]
Wrought Iron Dull	0.94	25	[84]
Wrought Iron Dull	0.94	349	[84]
Wrought Iron Polished	0.28	38	[84]
Wrought Iron Smooth	0.35	38	[84]
Zinc Bright Galvanized	0.23	38	[84]
Zinc Commercial 99.1%	0.05	260	[84]
Zinc Galvanized	0.28	38	[84]
Zinc Galvanized Sheet Iron, Fairly bright	0.230	28	[85]
Zinc Highly Polished	0.04–0.05	200–300	[85]
Zinc oxidized	0.280	24	[85]
Zinc Oxidized	0.11	260–538	[84]
Zinc Oxidized by heating at 399°C	0.110	399	[85]
Zinc Polished	0.02	38	[84]
Zinc Polished	0.03	260	[84]
Zinc Polished	0.04	538	[84]
Zinc Polished	0.06	1093	[84]
Zinc Unoxidized	0.05	300	[85]
Zinc, Galvanized Sheet	0.210	100	[85]
Zinc, Sheet	0.20	0°C	[87]
Zinc polished	0.045	300 K	[86]
Zinc Tarnished	0.25	300 K	[86]
Zirconium Silicate	0.920–0.800	238–500	[85]
Zirconium Silicate	0.800–0.520	500–832	[85]

Notes:
*Emissivities of almost all materials are measured at 0°C but do not differ significantly at room temperature.
**Paint, silver finish is measured at 25°C and Paint, enamel at 27°C
***When range of values for temperature and emissivity are given, end points correspond and linear interpolation of emissivity is acceptable.

Index

Printed in the United States
by Baker & Taylor Publisher Services

Printed in the United States
by Baker & Taylor Publisher Services